LASER RAMAN
GAS DIAGNOSTICS

A PROJECT SQUID WORKSHOP

LASER RAMAN GAS DIAGNOSTICS

Edited by

Marshall Lapp and C. M. Penney

Corporate Research and Development
General Electric Company
Schenectady, New York

PLENUM PRESS • NEW YORK AND LONDON

Library of Congress Cataloging in Publication Data

Project Squid Laser Raman Workshop on the Measurement of Gas Properties, Schenectady, N.Y., 1973.
Laser Raman gas diagnostics.

Sponsored by Project Squid, the Air Force Aero Propulsion Laboratory,
and the General Electric Co.
Includes bibliographical references.
1. Raman effect—Congresses. 2. Raman spectroscopy—Congresses. 3.
Gases—Analysis—Congresses. 4. Laser spectroscopy—Congresses. I. Lapp,
Marshall, ed. II. Penney, C. M., ed. III. Project Squid. IV. Air Force Aero
Propulsion Laboratory. V. General Electric Company. VI. Title.
QC454.R36P76   1973   535'.846   74-8800
ISBN 0-306-30800-2

Proceedings of the Project SQUID LASER RAMAN WORKSHOP
ON THE MEASUREMENT OF GAS PROPERTIES, held in
Schenectady, New York, May 10-11, 1973, and sponsored by
the Office of Naval Research (Project SQUID), the Air Force
Aero Propulsion Laboratory, and the General Electric Co.

© 1974 Plenum Press, New York
A Division of Plenum Publishing Corporation
227 West 17th Street, New York, N.Y. 10011

United Kingdom edition published by Plenum Press, London
A Division of Plenum Publishing Company, Ltd.
4a Lower John Street, London W1R 3PD, England

Printed in the United States of America

FOREWORD

The Laser Raman Workshop on the Measurement of Gas Properties is one of a series of occasional meetings organized in an informal workshop format through the stimulation of Project SQUID, Office of Naval Research. This workshop is the second to be organized on gas-phase applications of Raman scattering. Both Raman workshops were supported by Project SQUID, ONR, and the Air Force Aero Propulsion Laboratory, Wright-Patterson Air Force Base. The first Raman Workshop was held at the AVCO Everett Research Laboratory, Everett, Massachusetts, with their co-sponsorship in January 1972 under the chairmanship of D. A. Leonard. The present meeting was co-sponsored by the General Electric Research and Development Center, and held at their facility in Schenectady, New York. We are grateful to Project SQUID, AFAPL, and GE for their generous financial support of this Workshop, and to Project SQUID for underwriting the publication costs of the Proceedings.

As is always the case for successful meetings, many people contributed substantially to the organization and execution of this workshop. Professor Robert Goulard supported, aided, and encouraged us in the most helpful ways, and we are indebted to him. We received further valuable support and assistance from Dr. Ralph Roberts, Director, and Mr. James R. Patton, Jr., of the Power Branch, Office of Naval Research; from Dr. William H. Heiser, Chief Scientist of the Aero Propulsion Laboratory; and from Dr. James M. Lafferty, Manager of the Physics and Electrical Engineering Laboratory, and Dr. Donald R. White, Manager of the Optical Physics Branch, at the General Electric Research and Development Center.

On the first morning, Dr. White introduced the meeting, Dr. Lafferty gave a welcoming address, and Dr. Roberts gave some comments in a keynote speech, all in a session chaired by Dr. Heiser. We are indebted to these gentlemen for their enthusiastic sendoff.

The overall format of the Workshop involved about half the meeting being devoted to "formal" talks (i.e., prearranged, of 15 to 45 minutes duration each), with the other half of the meeting devoted to open discussion and short prepared or ad hoc presentations. Within this framework, the meeting was divided into four sessions:

May 10 - Morning   - Introduction; Density Measurements.
                  (Session Chairman, W.H. Heiser - AFAPL)

May 10 - Afternoon - Temperature Measurements; Chemistry.
                  (Session Chairman, A. B. Harvey - NRL)

May 11 - Morning   -   Resonance Effects; Remote Probes;
                           Experimental Advances.
                           (Session Chairman, W. Kiefer - Munich)

May 11 - Afternoon   -   Technology Applications.
                           (Session Chairman, D. Bershader - Stanford)

The "formal" talks were grouped into the first three sessions together
with discussion, while the fourth session was totally reserved for open
discussion and brief presentations. Note, however, in these Proceedings,
that some of these brief presentations have resulted in some considerably
expanded manuscripts, where the meeting discussion and author's requirements warranted it. In other cases, the brief manuscripts are intended
here to only outline the work described, and the authors intend the
readers to go to other publications for expanded discussion.

On the evening of May 10, following the conference dinner, we were
treated to a most enjoyable chamber music concert (which, depending upon
one's point of view, either stimulated our collective esthetic juices or
soothed our conference-worn savage breasts). The concert was organized
by our colleague (and flutist) Henri Fromageot, to whom we are grateful.

We are particularly delighted to acknowledge the conscientious and
tireless help of Ms. Nancy Cox (Purdue University) for all the secretarial
tasks associated with the meeting as well as the production of complete
audio tapes of all talks and discussion. Furthermore, Ms. Cox, together
with her co-workers, typed the entire Proceedings in a most expert fashion.

Local help in organizing and running the workshop was supplied by
many people, to whom we express our gratitude. Ms. Sandy Wroblewski (GE)
provided much assistance both preceding and during the meeting, and
Dr. Donald R. White not only helped with the meeting arrangements, but
also showed remarkable tolerance concerning all of the (anticipated but
not really expected) frustrations associated with hosting a conference.

Finally, we express our appreciation to all the attendees, who made
the Workshop a stimulating and worthwhile evaluation of Raman gas-phase
diagnostics.

                              Marshall Lapp - Chairman

                              C. M. Penney - Vice-Chairman

CONTENTS

INTRODUCTION

The purpose of the meeting reported in this Proceedings was to explore the application of Raman scattering techniques to the measurement of the state properties of gases, focusing upon fluid mechanic and combustion processes. The application of Raman scattering techniques to fluid mechanics and combustion technology is just now beginning. Concurrently, substantial interest has been shown in applying Raman scattering to a variety of other areas, such as meteorology, atmospheric pollution monitoring, exhaust gas analysis, industrial process control, chemical laser diagnostics, etc. While Raman scattering has for a great many years been a valuable laboratory tool in studies of the structure of molecules and the solid state, it has only recently come into prominence for the broader class of applications mentioned above. This prominence is due in part to the present availability of powerful, yet relatively inexpensive laser sources, and of high sensitivity photodetector systems. Furthermore, the rate of development of laser sources and photon detection apparatus is extremely high at the present time, so that increased usage of these scattering techniques will clearly exist over the next several years. We believe that the application of Raman scattering to fluid mechanics and combustion technology is not only a natural course of events in its own right, but that this application will also benefit from results obtained in a variety of other diagnostic applications of Raman scattering which are currently attracting large amounts of research effort.

The development of Raman scattering probe techniques is motivated both by the limitations of conventional probes, and by the many desirable characteristics of this process for probe applications. Some of these characteristics, which are brought out more clearly and suitably qualified in subsequent papers, are listed below.

Specificity. The spectral shift of Raman-scattered light is equal to a vibrational or rotational frequency of the observed molecule. Since these frequencies are different for different molecules, and change with their level of molecular excitation, it is possible to identify each line in a Raman spectrum with a particular type of molecule and its level of excitation.

Well-determined and independent response. The intensity of a Raman scattering line is directly proportional to the number density of corresponding molecules and independent of the density of other molecules. Experimental and theoretical evidence has supported these conclusions over wide ranges of gas pressure and temperature.

Three-dimensional resolution. On a "laboratory" scale, three-dimensional zones, with each dimension smaller than one millimeter, can be resolved. Furthermore, because Raman scattering is effectively instantaneous, the time dependence of the observed scattering can be used to provide range resolution. This so-called lidar (light detection and ranging) technique, which is analogous to that employed in radar, has been especially useful on a "field" scale -- particularly, in atmospheric measurements.

"Instantaneous" time resolution. In favorable cases, information can be time-resolved to the nanosecond or even shorter time regimes using pulsed lasers. With suitable signal analysis this information can be displayed "on line" in "real time".

Non-perturbing. The coupling interactions between optical radiation and non-absorbing gases is extremely weak. Consequently, over a wide range of experimental conditions, the state of the observed gas is not perturbed by scattering probes. Exceptions occur only when the exciting light is strongly absorbed, and/or when a short, energetic pulse is focused tightly, producing a non-linear optical response or gas breakdown.

Remote in situ capability. In general, light-scattering measurements do not require hardware near the measurement point, and can be applied to any system allowing optical access of moderate to good quality.

Accessibility of temperature information. For gases in thermal equilibrium, the Raman spectrum depends on both density and temperature. However, the temperature dependence is independent of density, and is sufficiently strong in appropriate spectral regions to allow sensitive temperature measurements from cryogenic through combustion temperatures.

Accessibility of density information. Density information from Raman scattering is effectively independent of temperature at temperatures below that at which appreciable vibrational excitation occurs in the observed species. Even at higher temperatures, the temperature correction to an appropriate density measurement can be moderate, such that extremely accurate temperature measurements are not required for useful density information.

Capability to probe systems not in equilibrium. Information determining composition and levels of internal mode excitation is readily accessible in non-equilibrium systems.

Simultaneous multiplicity of information. Numerous different components in a gas can be observed simultaneously.

<u>Relative lack of interference</u>. The vibrational Raman lines of
many molecules are well separated. In some cases interferences
do exist; viz. CO is obscured by N_2 when the former is at very
low relative concentration. Signals from different complex
molecules often overlap, yet even in such cases a useful measure of constituency is possible from the intensity of a Raman
band arising from a common chemical bond.

<u>Wide sensitivity</u>. Each molecule has at least one allowed Raman
band, in contrast to infrared absorption, which is not sensitive
to molecules without a permanent dipole moment (such as N_2 and
O_2).

Standing in contrast to this array of generally advantageous characteristics, there is one outstanding disadvantage -- the weakness of the
Raman effect. Typically, Raman scattering cross sections are three orders
of magnitude smaller that Rayleigh scattering cross sections, and roughly
ten to sixteen orders of magnitude smaller than cross sections for
unquenched fluorescence. This weakness places demanding requirements on
equipment required to obtain Raman scattering measurements to very high
accuracy, to very fine spatial or temporal resolution, at low concentrations and/or at long range. These requirements are set forth clearly in
the first paper. However, results reported in this Proceedings indicate
that these requirements have already been met in a number of significant
cases for which the importance of the measurement justifies the necessary
care.

In the Workshop and this resultant Proceedings, emphasis has been
placed on the development of new techniques, considered both from the
point of view of the underlying basic physics and chemistry, and the
ultimate applications to provide useful measurements in gases -- particularly gases under highly stressed conditions, such as at elevated
temperatures. Some of the types of overall questions which we hoped to
address were:

What are relevant and important fluid mechanic problems for
which Raman scattering can contribute new information?

What are the directions that Raman workers should take in
developing diagnostic capabilities in order to couple more
closely to fluid mechanics needs?

What requirements of combustion are different from those of
fluid mechanics?

What non-thermal equilibrium problems are most important?

What are the apparatus limitations for advancing the state-
of-the-art? Lasers? Spectral discrimination systems?
Signal processing? ....

What are the future directions, as we currently view them, for
general research in Raman diagnostics of gases?

and so on.

Obviously, we did not expect all of these questions to be fully
answered at the Workshop and naturally, new questions were introduced.
To a large extent, the course of discussion was determined by the mix of
people present -- a nearly even grouping of technique-developers and
users. The exchange was evident throughout the two days of close interaction, but it was particularly noticeable in the last session, in which
extensive discussion developed around informal presentations.

In many cases, the scheduled talks, informal presentations and discussion have been altered somewhat from their original oral form in
transcription to these Proceedings. In particular, some of the authors
have taken the opportunity to add new information, with the editors'
encouragement. In the case of the discussion, the editors have rearranged
some comments in an attempt to smooth the flow of ideas. It is our hope
that any distortion or omission introduced by this procedure will be
interpreted with a liberal spirit.

A final comment is irresistible. In any collaboration between co-
editors, there is a chance that the feeling will develop that one has
supplied the glimpses of brilliance, whereas the other has addressed the
gritty details. Each of us wishes to express his appreciation to the
other for taking care of the more onerous task.

                              Marshall Lapp and C. M. Penney, Editors

SESSION I
INTRODUCTION; DENSITY MEASUREMENTS
Session Chairman: W. H. Heiser

LASER RAMAN SCATTERING APPLICATIONS

by

R. Goulard

Project SQUID
Purdue University

ABSTRACT

The advent of laser sources, better optical instruments, and photon counting techniques has put Raman scattering within the range of a number of practical uses. The measurement of densities and temperatures has been demonstrated in numerous laboratory and field situations. In this paper, we consider the factors that determine the feasibility of Raman scattering measurements. The compromise between accuracy and time resolution is discussed in terms of desirable applications to short-lived processes (turbulence, kinetics, etc.).

I. INTRODUCTION

Rapid advances in lasers, sensors, and optical systems have brought the Raman effect within reach of technological applications. The possibility to "beat" the individual frequencies of a variety of molecules (rotational or vibrational modes) against that of an illuminating laser offers great potential in the remote measurement of the properties of complex mixtures. From a user's viewpoint, it is necessary to try to anticipate those areas of application which are the most likely candidates for laser Raman systems. It is also important to try to establish what research and development directions are more likely to foster a broad applicability of such systems. Hence the enthusiastic support given by AFAPL and ONR to this timely workshop.

II. DEVELOPMENT OF FEASIBILITY INDEX

In order to illustrate the areas where applicability seems feasible with present equipment and techniques, let us consider the basic equation giving the number E_s of scattered photons received by a laser Raman sensor (Figure 1):

3

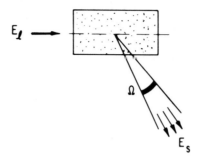

Figure 1.   Measurement geometry.

$$E_s = E_\ell \sigma_\ell (\nu_\ell / \nu_s) \, NL\Omega e \tag{1}$$

where:

E_ℓ is the number of photons from the laser source
          illuminating the volume of gas under observation

L   is the length of this sample in the direction of
     the laser beam, in cm

N   is the number density of the particles of interest,
     in cm^{-3}

σ is the Raman scattering cross section at the laser
     frequency, in $cm^2 \, sr^{-1}$

Ω is the solid angle subtended by the receiving area
     of the sensor, in sr

e   is the optical efficiency of the system

ν_ℓ is the laser frequency

ν_s is the scattered light frequency.

The ratio (ν_ℓ/ν_s) appears in Eq (1) because we use the usual light scattering cross section defined in terms of light energy rather than number of photons.

It is convenient to regroup the terms in Eq (1) so that the terms on the left-hand side relate to the laser-detection system, and those on the right-hand side to the geometry and nature of the experiment. In order to separate out the fourth power frequency dependence of Raman cross section on light frequency, let

$$\sigma = \sigma_0/q^4 \qquad\qquad (2)$$

where σ_0 is the cross section at a convenient reference frequency ν_0, and

$$q \simeq (\nu_0/\nu_\ell). \qquad\qquad (3)$$

Then, neglecting the ratio $(\nu_\ell/\nu_s) \simeq 1$, Eq (1) can be written in the form

$$q^4 \frac{E_s}{E_0} = NL\sigma_0\Omega e \qquad\qquad (4)$$

The expression on the right-hand side of Eq (4) can be regarded as a "feasibility index" X; i.e.

$$X \equiv NL\sigma_0\Omega e \qquad\qquad (5)$$

III. APPLICATION OF FEASIBILITY INDEX

The feasibility index is tabulated for four different kinds of applications (Refs. 1-6) in Table I. The reference frequency ν_0 has been chosen to correspond to a wavelength of 488 nm, since many Raman cross sections are available at this frequency. Table I indicates that density measurements at close range and near STP $(X \simeq 10^{-13})$ appear to be the most feasible of those considered, whereas long range pollution sensing (low ppm densities) appears more difficult $(X \simeq 10^{-19})$.

In order to gain some idea about how laser and detector capabilities match against these coefficients of feasibility, the left-hand side of Eq (2) can be specified in terms of the type of laser employed, and the required measurement accuracy. Suppose the required accuracy is ±10%. Statistical theory[2,7] indicates that at least 100 photons must be detected in order to achieve this accuracy. Thus we find that the minimum value of the feasibility index at which a satisfactory measurement can be obtained is

$$X_{min} = 100 \ q^4/E_{\ell}\eta \tag{6}$$

where η is the quantum efficiency of the detector at the scattered
light frequency. Values of X_{min} are tabulated in Table II for three

Table I

Feasibility Index X for Several Applications

	Refer.	L, cm	N, cm^{-3}	$\sigma_0, \dfrac{cm^2}{ster}$	Ω, ster	e	X
Density near STP (laboratory), vib.	1	1	2.5×10^{19}	5.4×10^{-31}	0.1	0.1	1×10^{-13}
Temperature (laboratory), rot. or Q branch	2,3	1	2.5×10^{19}	4×10^{-31}	0.1	0.1	1×10^{-13}
Atmospheric water vapor (field meas.), vib.	4	3×10^3	2.5×10^{18}	1.5×10^{-30}	10^{-6}	0.1	1×10^{-15}
SO$_2$ Pollutants-10 ppm (field meas.), vib.	5,6	3×10^3	2.5×10^{14}	2.5×10^{-30}	10^{-6}	0.1	2×10^{-19}

Table II

Values of X_{min} for $\pm10\%$ Accuracy

	n, sec^{-1}	q^4	η	E_{ℓ}	X_{min}
N$_2$ Pulsed Laser λ = 3371 Å; 1 mJ	500	0.228	20%	1.7×10^{15} per pulse	6.7×10^{-14} (single pulse)
Ar$^+$ Continuous Laser λ = 4880 Å; 1 watt	-	1	20%	2.46×10^{18} sec^{-1}	$2.03\times10^{-16} x\frac{1}{t}$
Ruby Pulsed Laser λ = 6943 Å; 1 J	10^{-1}	4.04	10%	3.51×10^{18} per pulse	1.15×10^{-15} (single pulse)

presently available laser-sensor combinations: two pulsed lasers and one cw laser. Values of X_{min} for $\pm1\%$ accuracy are larger by a factor of 10^2, since for this accuracy 10^4 photons must be detected.

Figure 2 illustrates the values of X_{min} corresponding to these three combinations. The vertical line corresponds, for each pulsed laser, to the measurement geometry (index X) that allows 10% accuracy with one pulse only. On the right of this line, more photons would be scattered into the sensor and a better signal-to-noise ratio would be obtained (better than 10% accuracy). On the left side of the vertical line, less than 10% would be obtained.

Note the apparent superiority of the 1J ruby laser over the 1mJ nitrogen laser when it is necessary or convenient to effect a measurement with one pulse only (i.e. in roughly 20 nanoseconds). On the other hand, Figure 2 also illustrates the use of pulsed lasers in a repetitive mode, in which case the more time is available, the more photons are scattered into the sensor, allowing even for low feasibility situations ($X \simeq 10^{-20}$, $t \simeq 20$ minutes). Note in this case, the relative advantage of the nitrogen laser over the ruby laser, due to the much higher pulse rate of the nitrogen laser. (This situation does not, however, hold true for probing luminous sources, for which high energy-per-pulse lasers are desirable to overcome the effect of background radiation.)

Figure 3 is based partially on the values in Tables II and III. It

Table III

Characteristic Time and Length for Some Measurement Applications (see Fig. 3)

	Characteristic time t, sec	Characteristic length L, cm
A. Jet Engine Compressor Inlet	10^{-3}	1
B. Compressor Blade Flow	$10^{-5} - 10^{-4}$	$10^{-1} - 1$
C. Flame Temperature Measurements		
1. quiescent	10	10
2. unsteady	10^{-1}	10^{-1}
D. Major Atmospheric Constituents	$1 - 10$	$10^2 - 10^4$
E. Small Scale Turbulence (high speed flow)	$10^{-5} - 10^{-4}$	$10^{-1} - 1$
F. Pollution Concentrations	$1 - 10^2$	$10 - 10^3$

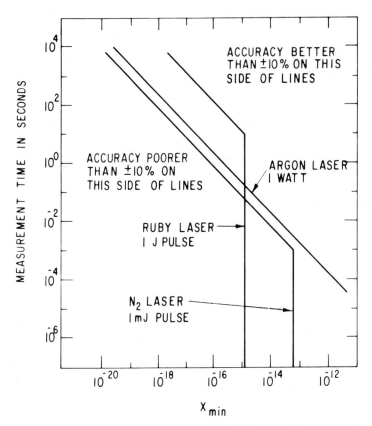

Figure 2.   Laser-detector system capability.   Values of X_{min}
for measurements to ±10% accuracy.

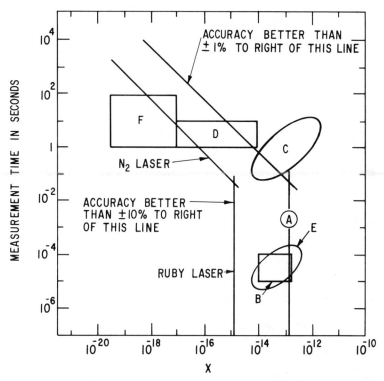

Figure 3.   Feasibility index X for several measurement applications.
Key: A - Compressor inlet. B - Compressor blade passage.
C - Laboratory temperature. D - Remote atmospheric constituents and temperature. E - Small scale turbulence
in laboratory. F - Remote pollution sources.

        For comparison purposes, the measurement potential of
current systems is illustrated for 2 accuracy figures.
Notice that the N_2 laser was chosen as typical of continuous (repeated pulse) measurement and the Ruby laser as
typical of one-pulse measurement.

        The diagonal appearance of cases C and E reflect the
observation that larger turbulence scales correspond usually
to longer characteristic times.

shows typical values of X corresponding to some possible applications of laser Raman techniques. Parts of the 10% and 1% accuracy lines of both N_2 and ruby for X_{min} are also shown to illustrate the measurement capabilities of the systems we have considered. It should be noted that the analysis we have based on Eq (4) does not apply directly to temperature measurements by Raman scattering, which are functionally dependent upon ratios of counts. However, the additional factors involved in temperature error calculations compensate somewhat in many cases,[7] so that our analysis gives a rough estimate for these measurements also.

The cases illustrated show the importance of the characteristic time t of the phenomenon under observation (see Table III). At the present time, it appears from Figure 3 that it is not possible to make accurate measurements of fluctuating properties with time scales less than one second, if these measurements must be obtained at long range (e.g. one mile away: cases D and F). Such were the field measurements discussed in Refs. 4 and 5, which necessitated several minutes of observation. Note also that if a time history of the measured N is needed, a continuous mode is necessary. Cases B and E illustrate two cases where a continuous measurement is not possible with presently available equipment.

IV. CONCLUSION

The challenge before us is one of uprating the capability of laser Raman systems and techniques, so that lower levels of feasibility X will become possible. It is possible to move both sections of the X_{min} curve to the left by using more powerful pulses or better quantum efficiencies. Also of significant interest is the possibility of a smaller q^4 in Eq (2): this could be accomplished by moving the laser frequency to higher values, which also produce scattering closer to resonance conditions. Hence the promise of doubling techniques and "near-resonance" scattering.[8,9]

Finally it should be kept in mind that 10% accuracy is not such a high standard of achievement, and that more detailed situations will usually involve smaller t's and L's (hence X's) than illustrated on Figure 3. Thus there is a need not only to improve laser Raman devices, but also to compare them[3,6,10] realistically to other measurement techniques (thermocouples, hot wires, electron beams, etc...). I wish you the best of luck in this difficult but promising task.

REFERENCES

1.   D. A. Leonard, "Point Measurement of Density by Laser Raman Scattering," AVCO-Everett Research Laboratory, Research Note 913 (1972); also, Project SQUID, Office of Naval Research, Technical Report AVCO-1-PU.

2.   J. A. Salzman, W. J. Masica, and Thom A. Coney, "Determination of Gas Temperatures from Laser Raman Scattering," NASA TN D-6336 (1971).

3. M. Lapp, L. M. Goldman, and C. M. Penney, Science 175, 1112 (1972);
    also, AIAA Paper 71-1084. See also M. Lapp, C. H. Penney, and
    R. L. St. Peters, "Laser Raman Probe for Flame Temperature," General
    Electric Research and Development Center, Project SQUID, ONR, Tech-
    nical Report GE-1-PU (April 1973).

4. S. H. Melfi, Applied Optics 11, 1605 (1972).

5. H. Inaba and T. Kobayasi, Nature 224, 170 (1969).

6. D. A. Leonard, "Development of a Laser Raman Aircraft Turbine Engine
    Exhaust Emissions Measurement System," AVCO-Everett Research Laboratory,
    Research Note 914 (1972).

7. M. Lapp, C. M. Penney, and J. A. Asher, "Applications of Light-
    Scattering Techniques for Measurements of Density, Temperature
    and Velocity in Gasdynamics," Aerospace Research Laboratories,
    Wright-Patterson Air Force Base, Report No. ARL 73-0045 (1973).

8. Ya. S. Bobovich, "Laser Spectroscopy Utilizing Spontaneous Raman
    Scattering of Weakly Interacting Molecules and Its Applications,"
    Soviet Physics, Uspekhi 15, No. 6, pp. 671-687, May-June 1973
    (Russian original 108, No. 3 and 4, Nov-Dec. 1972).

9. S. S. Penner and T. Jersky, "Use of Lasers for Local Measurement of
    Velocity Components, Species Densities and Temperatures," Annual
    Review of Fluid Mechanics, Vol. 5 (Annual Reviews Inc., Palo Alto,
    California, 1973), pp. 8-30.

10. S. Lederman and J. Bornstein, "Temperature and Concentration
    Measurements on an Axisymmetric Jet and Flame," Polytechnic
    Institute of New York, Project SQUID, ONR, Technical Report
    PIB-32-PU (Dec. 1973).

GOULARD DISCUSSION

HEISER - Surely after all those remarks there are questions. But
first I have one comment. To know thermocouples is not necessarily to
love them. True, they are comparatively simple, and have useful high
frequency response. But they can't stand the environments that we like
to work in. Either high temperature or debris can wipe out a thermocouple
in short order, and they have been pushed about as far as they can go.
The advent of Raman scattering diagnostics reminds me of a possible close
analogy: once the electric light bulb was invented no amount of R & D
in the candle industry could have saved it.

HENDRA - On the subject of alternatives, what about using Rayleigh
scattering for density measurements? The Raman effect is very weak,
as I am sure most of us here know. Rayleigh scattering is about 1,000
times stronger. It can be separated from Tyndall scattering (from dust
particles) by filtering out the central component, using a single mode
laser with an interferometer or a filter composed of a vapor that absorbs

in conveniently placed narrow lines, such as iodine. On the subject of
temperature measurements, I was under the impression that high temperature
measurements are best measured using emission processes because they
provide plenty of light. I believe that this approach is much easier than
the Raman method and therefore preferable unless there is a real reason
to use the Raman method (and, of course, sometimes there is). So, I
would suggest as a reply to your address that we be very careful not to
think that Raman scattering is the answer to all of our measurement
problems, but rather that it gives us new data. Let me give you an
example; nobody would normally measure the distance and direction of a
smoke cloud on the horizon by the Raman method. They would use lidar.
This is a pretty well-established technique. Frequently one does not
want to know what is in the smoke. Smoke is smoke, and if the law says
smoke is against the law, maybe that's enough. On the other hand, if
you really do need to know what is in the smoke, then perhaps the Raman
effect is worth considering. So let us not be too sophisticated. I
think we should stick to methods that are technologically easier than
Raman scattering, if possible, to make the measurements which you mention
in your talk.

   LEONARD - The comment by Hendra on Rayleigh scattering raises a
question. Granted that Rayleigh scattering is about three orders of
magnitude stronger than Raman scattering, I was wondering just how far
down on the Rayleigh wing one must go to avoid detecting direct laser
scattering from dirt and walls, and at that point on the wing how large
would the remaining Rayleigh signal be? We need to answer this question
before a choice can be made between Rayleigh and Raman scattering.

   PENNEY - I couldn't agree more with Pat Hendra's remark to the
effect that Raman scattering isn't sacred. However, we have looked
fairly carefully at the possibility of density measurements by Rayleigh
scattering. These measurements may be very useful in systems where
the gas composition is known, and particle scattering is insignificant
or can be evaluated from the time dependence of the signal. But,
whenever narrow line filters must be used to screen out strong particle
scattering, there can be serious problems. For example, the source
laser must then provide an extremely narrow and stable spectral line,
which is expensive and reduces laser power. Second, one needs to know
the fraction of the Rayleigh peak that is blocked by the filter. This
fraction depends on composition and temperature. It also depends on how
well the peak follows the Gaussian distribution of Doppler broadening
predicted for an ideal gas. Even in backscattering, Brillouin peaks
may begin to appear at high gas density. Third, very narrow line
filters with large rejection ratios are not easy to obtain or use.
It appears that Rayleigh scattering may be very useful for measurements
under appropriate conditions, but these conditions may be fairly
restrictive.

   I do want to mention one other subject that was raised in this talk.
There are a number of people who would like to make density or temperature measurements in a microsecond or less in order to look at turbulence
or time correlation in gas flows. One way to get this time resolution
is to use very high energy pulses. However, a limit is reached in this
approach when the pulse breaks down the gas. Another way is to use lower

energy pulses which come in pairs, separated in time by microseconds
to milliseconds. The information from a large number of these pulse
pairs can be collected to reduce statistical errors, and combined to
provide density or temperature correlation functions.

LAPP - I would like to point out that there are people here from
many disciplines, including a wide variety of aerodynamic and gasdynamic
backgrounds, and an equally wide variety of spectroscopic and diagnostic
backgrounds. Comparison of the various diagnostic techniques for the great
variety of gas-phase measurement needs is a major goal of this workshop
and the mix of people present is a good one for this task.

One further comment: one way of detecting more scattered light is
through the use of the multiplex interferometric technique that Joe
Barrett has developed for rotational Raman scattering. He will be describing this technique later this morning.

LEONARD - Before you sit down, Marshall, what has been your experience
with the problems of sodium line reversal temperature measurements?

LAPP - The problem is that sodium line reversal provides column
rather than point measurements and thus it is subject to strong complications arising from gradients in the properties of the test system.
Self absorption and cold boundary layers are often particularly difficult to handle in data reduction. If you assume simple forms of symmetry,
such as cylindrical, temperature results can be obtained with mathematical
inversion techniques. However, several people here have worked very hard
for years obtaining this kind of information, and the problems are often
formidable.

BERSHADER - I have come here to learn about a topic in which I haven't
worked, but it seems to me it would be desirable to get a handle on all
the different parameters which have been mentioned here. Maybe a good bull
session is the way to do it. One would like to devise something like a
figure of merit that is used for other instrumentation. Firstly, there
is the cross section for the scattering phenomena. Next, the matter of
time resolution is clearly of importance, especially with respect to
fluid dynamic and propulsive applications. Following that, we ask about
space resolution, i.e., the size of the scattering volume. Then there is
the old question of the interaction of the measuring device with the
system to be measured. Here, I would like to think in terms of the ratio
of focused laser energy in the scattering volume that is coupled by some
mechanism to the fluid, to a logically defined fluid energy (such as the
total enthalpy, the energy associated with turbulent fluctuations, etc.).
Other obvious factors include useful ranges of the concentration or
temperature variables and even such practical features as the versatility
of the Raman equipment in adaption to different kinds of test configurations. Is it possible to put the collective wisdom together ultimately
in a more organized fashion along these lines?

SCHILDKRAUT - I would just like to make a comment as a nonspectroscopist,
instrumentation specialist in engineering. Dr. Goulard has calculated a
feasibility coefficient. There is also a coefficient of "hairiness"
that is appropriate to these measurements. For example, we manufacture

interferometric infrared instrumentation; we would like to sell these
devices for use in air pollution measurements. However, when you aim
an IR instrument at a smokestack your goal is to quantify what is there.
After you get finished with the Beer's law calculations and account for
a dozen potential interferences, you still may not know what is coming
out of the stacks or how much.

Raman relative to IR is a second order effect and hence is weak
by comparison. But fortunately, the number of Raman photons that comes
back to your spectrometer is directly proportional to what is in the well-
defined beam, which makes the calculations so much easier. Going to
complex filters and dealing with other subtleties sometimes makes the
overall measurement accuracy more questionable. Nevertheless, in favor
of remote Raman spectroscopy is the ability of engineers like myself to
perform the measurements correctly without going through a lot of somewhat
questionable calculations on occasion. Making fewer assumptions and
getting the answer is what's relevant. Raman scattering may not be as
good as we can do; it is not a panacea - but it is viable and fieldable.

GOULARD - I would like to emphasize that the question of optimized
information treatment is critical, because the Raman effect is such a
weak one. Therefore, one must devise procedures which give maximum
accuracy from a minimum of well chosen data. Furthermore, this
procedure should be general enough to be feasible for a wide range
of field conditions.

WHITE - I would like to point out the way work on Raman scattering is
benefitting our parallel programs, and in particular, Air Force programs.
Murray Penney pointed out how one could use two pulses separated by a
small amount of time. The Air Force is now interested in developing
neodymium lasers that have exactly this capability; ten pairs of pulses
per second where the two pulses in each pair are separated by a small
and controllable time interval. With frequency doubling, this capability
could produce a very nice probe source for the use of Raman scattering
to study problems such as turbulence.

ANALYSIS OF RAMAN CONTOURS IN VIBRATION-ROTATION SPECTRA

by

R. Gaufres

Laboratoire de Spectroscopie Moléculaire
Université des Sciences et Techniques du Languedoc
Montpellier, France

ABSTRACT

As is widely known, a Raman band is a gas phase spectrum always exhibits a rotational structure. Even a totally symmetric band of a spherical top, which consists only of a Q-branch, presents a profile which may be calculated in terms of variation of rotational energy, from a lower vibrational state to an upper one.

In most cases, the rotational structure of the Raman bands is not resolved, and one observes contours depending on (i) the various molecular constants entering the expression for the rotational energy, (ii) the temperature, (iii) the type of the rotator and the symmetry of the vibration, and (iv) the polarization of the scattered light ($I_{||}$ or $I_{\perp}$).

Actually, a "band" may be a superposition of a system of bands, associated with several vibrational transitions of slightly different energies, such as a fundamental and the corresponding upper-state transitions (or "hot bands"). In this case, the observed contours depend, in addition to the various aforementioned parameters, on anharmonicity constants, and on the temperature, which is not necessarily the same as above, when the sample is out of thermodynamic equilibrium.

The a priori calculation of an unresolved Raman contour is of interest, as it allows either a measurement of molecular constants or a determination of temperature. In every situation, the possibility of a computer simulation of a Raman band does exist, but, in the simplest cases, approximate analytical expressions can be derived for the "natural" contours (unperturbed by the apparatus function) and also for the convoluted contours. The approximations entering these calculations are allowed when $Bhc \ll kT$. The analytical expressions enable one to make an easier analysis of the band contours.

When one is interested in temperature measurement, he may only consider the trace scattering ($I_{||} - 4/3\ I_{\perp}$), i.e. the trace contribution to the Q-branch of totally symmetric vibrations. The polarized Q-branches are generally strong, and the intensity distribution among the individual lines is given by a simple law. Moreover, in a system of bands, the contour of the trace-scattered light is best resolved.

The principles of calculation of the Q-branch profiles will be emphasized for the cases of linear molecules, symmetric tops, and spherical tops, and the case of the asymmetric top will be raised.

I. <u>FEATURES OF THE STRUCTURE OF A RAMAN VIBRATION-ROTATION BAND</u>

Quite generally, when a vibrational Raman transition occurs in the gas phase, the rotational energy varies, so that the vibrational Raman "lines" are actually a juxtaposition of many rotational lines on both sides of the frequency corresponding to the vibrational energy change. According to the type of rotator (linear, symmetric top, spherical top, asymmetric top), the rotational energy is given in terms of various sets of quantum numbers: J for the linear rotator and the spherical top, J and K for the symmetric top, J and a "pseudo quantum number" (for instance τ) for an assymetric top.[†] The rotational spectral terms may be written:

$$E/hc = F(J,n),\qquad\qquad(1)$$

where n represents the quantum number other than J, which is universal. The expression (1) contains as parameters the rotational constants of the molecule under consideration, and these depend on its mechanical properties, such as moments of inertia.

According to the symmetry of the rotator and the symmetry of the vibrational levels[§] involved in the transition, different selection rules for the rotational quantum numbers can exist, which then lead to different series of rotational Raman shifts

$$\Delta\tilde{\nu}_{rot} = F(J', n') - F(J", n");\qquad\qquad(2)$$

that is, to different band types.

[†]For every type of rotator, a quantum state is also defined by a magnetic quantum number M, which does not enter the energy expression when no field is applied to the molecule. We shall omit M when it is not required.

[§]Actually, one must consider the symmetry of the vibronic levels involved in the Raman transition, but for sake of simplicity, we assume that the molecule remains in a totally symmetric electronic state, which is an assumption generally valid.

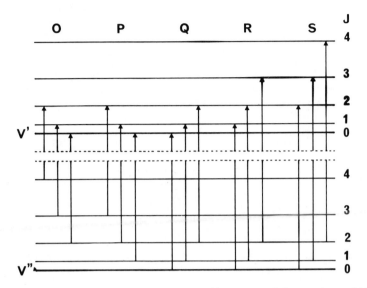

Figure 1.   Vibration-rotation energy diagram and Raman transitions,
with O, P, Q, R and S branches. The relatively simple
diagram is appropriate for a linear molecule or a
spherical top. The five branches occur in a $\Pi \leftarrow \Sigma$
band of a linear rotator ($\mathcal{C}_{\infty v}$) or in a E $\leftarrow$ A band of a
spherical top ($\mathcal{C}_d$ or O_h). The total energy change
is increasing from the left to the right.

    The most liberal selection rule on J in the Raman effect is
ΔJ = -2, -1, 0, 1, and 2, giving rise, respectively, to five series
of lines, the so-called O, P, Q, R, and S-branches. In Figure 1,
we see a diagrammatic representation of the Raman transitions for cases
where the rotational energies are defined by J only.

    On the other hand, the intensities of these individual lines in a
band are not equal. They depend obviously on the population of the
rotational level in the initial state, which is a function of the
temperature according to Boltzmann's law, and also on matrix elements
containing the rotational quantum numbers of the initial and final
states.

    The ensemble of all these lines gives rise to an intensity
distribution which may be resolved in the most favorable cases, but most
often is observed as a continuous smooth function of the frequency,
namely a Raman contour.

    The theoretical background of the rotational structure of the Raman
bands was developed in 1933 by Placzek and Teller.[1] Their conclusions
can be found in a more modern and comprehensive form in the books of

Herzberg.[2,3] Stoicheff[4] also gives an introduction to the Raman effect
in the gas phase. Until the development of the laser, no extensive
experimental work was devoted to the analysis of the Raman bands in the
gas phase, with the exception of high resolution studies of the spectra
of small molecules, mainly by the Canadian school. Nowadays, it is
possible with commercial devices to record "unresolved" Raman contours
with sufficient accuracy to allow a rigorous analysis. This paper is
devoted to such studies.

II.   TRACE AND ANISOTROPY SCATTERING

Since the scattering of light by molecules is governed by the
molecular polarizability, which is a second-order tensor property, the
symmetry behaviour of this phenomenon is more complex than that of the
absorption of light (which is due to a vector property, the dipole
moment of the molecule). The experimental information obtained from
the spectra can be, in turn, more extensive.

Clasically, the scattering of light can be understood in terms of
the radiation of the dipole induced by an alternating external field.
The three components of this dipole are written:

$$M_j = \sum_k \alpha_{jk} E_k \qquad (j, k = x, y, z) ,\qquad (3)$$

in which E_k is the k-component of the external field and the α_{jk} are the
elements of the polarizability tensor of the molecule, which is generally
symmetric ($\alpha_{jk} = \alpha_{kj}$).

Since the polarizability is a molecular property, the α_{jk} elements
have constant values when x, y and z are defined as an axis system
fixed to the molecule - the so-called molecular axes - and when the
molecule is in a vibrationless state.

The vibrational Raman scattering is explained in terms of the modulation of the polarizability elements by the vibration of the molecule.
The pure rotational scattering results from the modulation of the
elements of the polarizability tensor, when expressed in the laboratory
axis system, as the molecule rotates. We can write:

$$M_\rho = \sum_\sigma \alpha_{\rho\sigma} E_\sigma \qquad (\rho, \sigma = X, Y, Z) ,\qquad (4)$$

and

$$\alpha_{\rho\sigma} = \sum_{j,k} \cos(j,\rho) \cos(k,\sigma)\alpha_{jk} ,\qquad (5)$$

in which X, Y and Z define the laboratory axis system, and the $\cos(j,\rho)$ and $\cos(k,\sigma)$ are direction cosines. Then the rotational modulation arises through the changing direction cosines.

The vibration-rotation spectrum results from the combination of the two effects.

The quantum mechanical explanation of light scattering deals with matrix elements of the polarizability. To account for the vibrational spectrum, we consider matrix elements of the form

$$< \psi''_{vib} \; |\alpha_{jk}| \; \psi'_{vib} > ,$$

or, more simply

$$< v'' \; |\alpha_{jk}| \; v' > , \quad ^{*}$$

where the α_{jk}'s are defined in the molecular axis system, and considered as functions of the vibrational coordinates.

The pure rotational spectrum is explained in terms of matrix elements of the polarizability, referred to the laboratory axis system:

$$< \psi''_{rot} \; |\alpha_{\rho\sigma}| \; \psi'_{rot} >$$

or

$$< J'', n'', M'' \; |\alpha_{\rho\sigma}| \; J', n', M' > ,$$

which is a sum of terms of the form

$$< J'', n'', M'' \; |\alpha_{jk} \cos(j,\rho) \cos(k,\sigma)| \; J', n', M' > .$$

Finally, the probability of a vibration-rotation quantum jump can be calculated from matrix elements of the form

$$< v'', J'', n'', M'' \; |\alpha_{jk} \cos(j,\rho) \cos(k,\sigma)| \; v', J', n', M' > ,$$

---

*In this notation, v' and v'' must be thought as complete series of vibrational quantum numbers in the case of a polyatomic molecule.

in which the vibrational and the rotational coordinates can be separate and appear as products:

$$< v'' \left| \alpha_{jk} \right| v' > \; < J'', n'', M'' \left| \Phi_{\rho,\sigma,j,k}(\theta,\phi,\chi) \right| J', n', M' > .$$

Here $\Phi_{\rho,\sigma,j,k}(\theta,\phi,\chi)$ is a function of the Eulerian angles defining the orientation of the molecule in the laboratory axis system, ρ and σ refer to the directions of polarization of the scattered and exciting light, respectively, according to Eq (4), and j and k refer to the molecular polarizability element involved in the transition.

The intensity of a vibration-rotation Raman line, in an optical arrangement allowing only one direction of polarization (σ) for the exciting beam, and only one (ρ) for the observed scattered light, is proportional to the sum of the squares of such elements:

$$I_{\rho\sigma} (v',J',n',M' \leftarrow v'',J'',n'',M'') \propto \sum_{j,k} \left| < v'' \left| \alpha_{jk} \right| v' > \right|^2$$

$$\left| < J'',n'',M'' \left| \Phi_{\rho,\sigma,j,k} \right| J',n',M' > \right|^2 . \qquad (6)$$

A band is allowed when at least one of the six different elements $< v'' \left| \alpha_{jk} \right| v' >$ is non vanishing. The rotational structure of an allowed band arises from the non-vanishing elements of the second term of the product.

Actually, the polarizability tensor may always be written as the sum of two tensors, a spherical and a traceless one

$$\begin{vmatrix} \alpha_{xx} & \alpha_{xy} & \alpha_{xz} \\ \alpha_{yx} & \alpha_{yy} & \alpha_{yz} \\ \alpha_{zx} & \alpha_{zy} & \alpha_{zz} \end{vmatrix} = \begin{vmatrix} \alpha & 0 & 0 \\ 0 & \alpha & 0 \\ 0 & 0 & \alpha \end{vmatrix} + \begin{vmatrix} \alpha^1_{xx} & \alpha^1_{xy} & \alpha^1_{xz} \\ \alpha^1_{yx} & \alpha^1_{yy} & \alpha^1_{yz} \\ \alpha^1_{zx} & \alpha^1_{zy} & \alpha^1_{zz} \end{vmatrix} \qquad (7)$$

in which $\alpha = (1/3)(\alpha_{xx} + \alpha_{yy} + \alpha_{zz})$ is the mean polarizability,

$$\alpha^1_{jj} = \alpha_{jj} - \alpha,$$

and

$$\alpha^1_{jk} = \alpha_{jk} \; (j \neq k) .$$

The trace of the first term of the sum given on the right-hand side of Eq (7) is equal to the trace of the complete polarizability tensor. Its anisotropy is obviously zero. The reverse holds for the second term.

The same decomposition is also possible for the matrix elements entering Eq (6), namely

$$| < v" |\alpha_{jk}| v' |^2 \quad | < J",n",M" | \Phi_{\rho,\sigma,j,k} | J',n',M' > |^2 =$$

$$| < v" |\alpha \delta_{jk}| v' > |^2 \quad | < J",n",M" | \Phi^0_{\rho,\sigma,j,k} \delta_{jk} | J',n',M' > |^2$$

$$+ < v" |\alpha^1_{jk}| v' > |^2 \quad | < J",n",M" | \Phi^1_{\rho,\sigma,j,k} | J',n',M' > |^2 , \qquad (8)$$

and therefore, the intensity of a Raman line can be considered as the sum of two terms, one arising from the spherical part of the polarizability tensor, the trace scattering, and the other from the traceless part, the anisotropy scattering. These considerations enable several further statements to be made on the selection rules:

i) In the trace scattering term, the vibrational factor

$$< v" |\alpha| v' >$$

is zero unless v" and v' are of the same symmetry species, due to the symmetry of α. Roughly speaking, trace scattering occurs only for totally symmetric vibrations.

ii) As the orientation of the spherical part of the polarization is meaningless, it is obvious that

$$\Phi^0_{\rho,\sigma,j,k} \delta_{jk} = \Phi^0_{\rho,\sigma,j,k} \delta_{\rho\sigma} \delta_{jk} = \delta_{\rho\sigma} \delta_{jk}$$

and we can state that the only rotational transitions allowed in trace scattering are those for which

$$J' = J", \quad n' = n" \quad \text{and} \quad M' = M"$$

Thus, trace scattering can only be present in the Q branch ($\Delta J = 0$).

iii) Whenever the matrix elements of the spherical tensor are non vanishing, the corresponding matrix elements of the traceless tensor are also non vanishing, as they have the same symmetry behavior, unless $\alpha_{jj} - \alpha = 0$ (which occurs for the spherical top). Disregarding this case, we can state that anisotropy scattering is always superimposed to trace scattering, when the latter is allowed.

iv) All other Raman transitions, particularly those in which a rotational quantum jump occurs, are due to the off-diagonal elements of the traceless tensor. The anisotropy scattering is exclusively accountable for the O, P, R and S branches of a band.

Fortunately, there is a possibility to separate experimentally the two different parts of the spectrum. Let us consider a sample irradiated by a polarized beam (a laser beam for instance), and the light scattered at right angle, in a plane perpendicular to the electric vector of the polarized incident beam (Figure 2). As the scattering process generally has no revolution symmetry, the scattered light is polarized; the intensities of the two principal components are designated by $I_{\parallel}$ and $I_{\perp}$ for the two directions of the electric vector, parallel and perpendicular, respectively, to the electric vector of the incident light.

The quantities $I_{\parallel}$ and $I_{\perp}$ may be written as functions of the intensities of the trace and anisotropy scattering:

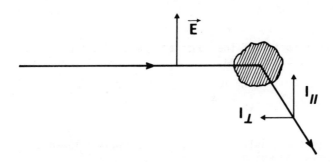

Figure 2.   Geometry of the scattering phenomenon. If $\vec{E}$ were lying in the plane defined by the two beams, the intensities of both components should be equal to $I_{\perp}$.

$$I_{\parallel} \propto 45\,I_{tr.} + 4\,I_{anis.}\ , \qquad\qquad (9a)$$

$$I_{\perp} \propto 3\,I_{anis.}\ . \qquad\qquad (9b)$$

When the geometry of the experimental arrangement is that of Figure 2, the intensity distribution is given by $I_{\perp}$ for the anisotropy scattering, and by $I_{\parallel} - 4/3\,I_{\perp}$ for the trace scattering. From the fundamental expressions (9a) and (9b), it is easy to derive the intensities of the observed components as functions of $I_{tr.}$ and $I_{anis.}$ when different optical arrangements are involved.

III. THE DIFFERENT RAMAN BAND TYPES

We shall briefly discuss now the different band types for each sort of molecular rotator. In what follows, we assume that the lower vibrational level is always totally symmetric.

A. Linear Molecules

The rotational spectral terms are given by

$$F(J) = B\,J(J+1) - D\,J^2\,(J+1)^2\ , \qquad\qquad (10)$$

where B and D are rotational constants; B is proportional to the reciprocal of the inertia moment and D accounts for the centrifugal effect. It is weak in comparison to B, and may be neglected in many cases. The degeneracy of each level is $2J + 1$.

There are two types of vibration-rotation band, according to the symmetry of the excited vibration.

1) Parallel bands

The upper vibrational level is Σ^{+} ($\mathscr{C}_{\infty v}$ symmetry group) or Σ_g^{+} ($\mathscr{D}_{\infty h}$ group). In this case, $\Delta J = 0, \pm 2$. There is a strong Q-branch, containing trace and anisotropy scattering, and two rotational wings, arising from the O and S-branches.

2) Perpendicular bands

The upper vibrational state is π or π_g. Then, $\Delta J = 0, \pm 1, \pm 2$. The Q-branch, containing only anisotropy scattering, is very weak and is

not observed, quite generally. The two rotational wings, formed by the
O,P and R,S-branches respectively, are the prominent features of the
band.

B. Symmetric Tops

The rotational spectral terms are given, for an oblate symmetric
top, by

$$F(J, K) = B\,J(J+1) + (C-B)\,K^2.^* \qquad\qquad (11)$$

For a prolate top, C must be replaced by A. Note that B and C (or A), are,
as in Eq (10), rotational constants proportional to the reciprocal of each
of the two different principal moments of inertia. The quantum number K
runs from $-J$ to J. The degeneracy of each level is $2(2J+1)$ for
$K \neq 0$, and $2J+1$ for $K = 0$.

1) Symmetric bands

The upper vibrational level is totally symmetric. Then $\Delta J = 0$,
± 1, ± 2 and $\Delta K = 0$. The Q-branch is strong (trace and anisotropy
scattering) and the rotational wings, formed by the O,P and R,S-branches,
are generally of poor intensity in comparison to the Q-branch.

2) E-type bands

When the upper vibrational level is doubly degenerate (E symmetry
species), different situations may occur, according to the symmetry
group of the molecule. In all cases, $\Delta J = 0$, ± 1, ± 2, but there are
E-bands with $\Delta K = \pm 1$ and ± 2, for instance in the spectra of $\mathscr{C}_{3v}$
molecules, E-bands with $\Delta K = \pm 1$, and others with $\Delta K = \pm 2$, in the
spectrum of benzene for instance. In molecules with a fourfold axis,
there is only one type of E-band, with $\Delta K = \pm 1$. The structure of these
bands is far more complex than all we have previously seen: in every
branch there may be two lines $(\Delta|K| = 1$ and $\Delta|K| = 2)$ for each combina-
tion of J and $|K|$ values. The Q-branch, the intensity of which is of
the same order of magnitude as the other branches, spreads over a
large spectral range and is no longer a prominent feature of the band.
Moreover, these bands may be strongly perturbed by first order Coriolis
interactions between vibration and rotation.

---

*A more complete expression, containing, such as in Eq (10), centrifugal
 parameters of lower order of magnitude than B and C, must be used when
 high accuracy is desired.

3) B-type bands

In molecules with a fourfold axis, besides the E-bands with $\Delta K = \pm 1$, there are also B-type bands, with $\Delta J = 0, \pm 1, \pm 2$ and $\Delta K = \pm 2$. The complexity is the same as for the case just discussed, although they are not perturbed by Coriolis interactions.

C. Spherical Tops

The rotational spectral terms are given by

$$F(J) = B J(J+1), \tag{12}$$

as is the case for a linear molecule, but the degeneracy is $(2J+1)^2$. There are three types of Raman bands according to the symmetry of the upper vibrational level, $A_{(g)}$, $E_{(g)}$, or $F_{2(g)}$,[*]

1) A-type bands

The selection rule is $\Delta J = 0$, so that the band consists only of a Q-branch, entirely due to trace scattering.

2) E-type bands

The selection rule is $\Delta J = 0, \pm 1, \pm 2$ and the band exhibits five branches of simple structure. The overall intensities of the five branches are equal. These bands are due to anisotropy scattering only.

3) F_2-type bands

The selection rule is the same as above, but the situation is more complex as each of the rotational levels in the excited vibrational state are split into three components by Coriolis interactions and each branch is the superposition of three sub-branches.

D. Asymmetric Tops

There is no simple expression for the rotational spectral terms of the asymmetric top, and a detailed statement concerning this subject

---

[*] The symmetry species are A, E and F_2 for $\mathscr{T}_d$ molecules, A_g, E_g and F_{2g} for $\mathscr{O}_h$ molecules.

is beyond the scope of this paper. For every J value, there is $(2J+1)$ levels designated as J_τ (τ running from $-J$ to $+J$). The spectral terms may be written, for instance, as

$$F(J_\tau) = \frac{A+C}{2} J(J+1) + \frac{A-C}{2} E_\tau ,$$
(13)

where the E_τ are the solutions of a set of equations in which the three rotational constants A, B and C, corresponding to the three principal inertia moments, enter as parameters; there is such a set of equations for each J value. The $(2J+1)$ levels refering to a given J value have different symmetry properties. For each of them, the degeneracy is $(2J+1)$.

The selection rule on J is $\Delta J = 0, \pm1, \pm2$, but all the transitions obeying this rule are not allowed. There are further selection rules, which take into account the molecular symmetry group, the symmetry species of the upper vibrational state, and the symmetry properties of the individual J_τ levels. Nevertheless, the structure of the Raman bands is very complex. Two main types may be considered.

1) Symmetric bands

The excited vibration is totally symmetric. Then, the Q-branch contains trace scattering and may appear as a strong central "line".

2) Antisymmetric bands

All the five branches arising from anisotropy scattering exhibit near-by overall intensities, and the intensity distribution in the band is very smooth.

IV. INTENSITIES OF THE INDIVIDUAL LINES

In order to calculate the intensity distribution in a Raman band, one must first consider the intensities of the individual lines. A general expression for these is:

$$I_{\rho,\sigma} \propto (\tilde{v}_0 - \tilde{v}_{Ram})^4 \, g_I \, g_{rot} \sum_{j,k} |<v''|\alpha_{jk}|v'>|^2$$

$$|<J'',n'',M''|\Phi_{\rho,\sigma,j,k}|J',n',M'>|^2 \exp[-E_{rot}/kT],$$
(14)

using the formalism of Eq (6) of Sec. II. Here $\tilde{v}_0 - \tilde{v}_{Ram}$ is the absolute frequency of the Raman line, g_I is the nuclear spin degeneracy in the initial state, which needs to be considered in high resolution studies only, and g_{rot} is the degeneracy of the initial rotational level. The exponential is, of course, the rotational Boltzmann factor.

For a given vibrational transition, that is, a given Raman band, when several elements $< v'' |\alpha_{jk}| v' >$ vanish,[*] there are two possible situations according to the value of the corresponding rotational matrix elements.

1) For every j,k pair which allows the vibrational transition, the rotational matrix elements have the same values. These may be factored in Eq (14) and the sum $\sum_{j,k} |<v''|\alpha_{jk}|v'>|^2$ is a constant of the band. That is the most common situation.

2) The rotational matrix elements have different values, leading to different rotational selection rules. The band may then be considered as the superimposition of two (or more) bands.

These considerations allow us to drop the summation in Eq (14). Using now the results of the decomposition of the polarizability, as discussed in Sec. II, it becomes possible to consider the trace scattering part of the intensity of an individual line, and the anisotropy scattering part, omitting the vibrational term, which is a constant of the band.

$$I_{tr.} \propto (\tilde{\nu}_0 - \tilde{\nu}_{Ram.})^4 \, g_I \, g_{rot} \, \exp [-E_{rot}/kT] \qquad (15)$$

$$I_{anis.} \propto (\tilde{\nu}_0 - \tilde{\nu}_{Ram.})^4 \, g_I \, g_{rot} \, |<J'',n'',M''|\phi^1_{\rho,\sigma,j,k}|J',n',M'>|^2$$

$$\exp [-E_{rot}/kT] \qquad (16)$$

The rotational matrix elements in Eq (16) are functions of the quantum numbers defining the levels involved in the transition. For the symmetric top, their squares are the so-called $b^{J,K}_{J',K'}$ coefficients of Placzek and Teller.[1] For their values, see also Gaufres and Sportouch.[5]

The linear rotator may be considered as a limiting case of a symmetric top, with $K = 0$, and the rotational matrix elements are then

_____

[*] As $\alpha_{kj} = \alpha_{jk}$ and is of the same symmetry species, the two elements may be considered as identical, with the intensity given by Eq (14) being multiplied by the factor 2.

easily derived. For the spherical top, see Herranz and Stoicheff.[6] The anisotropy intensities for the asymmetric top apparently have not been calculated up to this date.

V.   THE DIFFERENT METHODS OF CALCULATION OF A RAMAN CONTOUR

The rotational structure of the Raman bands is generally unresolved; i.e. one observes a contour. We shall discuss now the different methods for the a priori calculation of these contours.

The frequencies of the rotational lines can be obtained as differences of spectral terms, using the selection rules given in the previous sections for each case. The intensities of the lines have been discussed in Sec. IV.

A. Computer Simulation

In every case, it is possible to obtain a Raman contour by dividing the spectral range of the band into narrow spacings, and adding the intensities of all the lines that appear in each spacing. Strictly speaking, there is an infinite number of lines in a band, but in every branch the intensity approaches zero with Boltzmann's factor and one can stop the run in a series of lines (a branch for example) when the intensity of a line becomes, for instance, one hundredth of the intensity of the strongest one in the series. The observed contour is always the convolution product of the "natural" contour with the slit function. To get an accurate simulation of the band, one must introduce the slit function, and, for the sake of simplicity, one can assume in this method a triangular function. Thus, the intensity of the lines occurring in a given spacing must be considered for all the spacings overlapped by the triangle. This refinement is of course unnecessary when the base of the experimental triangle is narrower than a calculation spacing.

This method is quite general. The most accurate expressions for the spectral terms and the intensities, taking into account all the refinements of the theory, can be used, so that the accuracy of the results depends only on the duration of the computer runs. This method has been used by several workers.[7,8,9]

B. Analytical Expressions

Considering the molecular constants and the temperature as parameters, the frequencies of the lines and their intensities both appear as functions of the rotational quantum numbers: viz. $\vartheta(J,n)$ and $I(J,n)$. In some favorable cases, it is possible to eliminate the quantum numbers between the two previous expressions and to obtain an analytical expression of the band contour as a function of the frequency $I(\nu)$. This method has been used by Placzek and Teller[1] who calculated the rotational wings

(O,P and R,S-branches) of a symmetric band of a symmetric top. More recently, it has been extended to other situations.[9,10,11,12,13]

As long as the mathematical expression obtained is not too complex, it is possible to work out the convolution product by a slit function, and that is facilitated when one assumes for the slit function a guassian shape.[11]

VI. Q-BRANCHES AND OVERALL CONTOUR ANALYSIS

It becomes now more and more obvious that there are two different fields in the analysis of the Raman bands, namely the study of the Q-branches, in the bands due to totally symmetric vibrations, and the analysis of the whole contours. Actually, the difference lies in i) the physical nature of the phenomena, ii) the experimental problems to be resolved, and iii) the kind of information obtained. It should be more correct to speak about trace intensity and anisotropy intensity analysis.

In the first case, the distribution may be very sharp, displaying a "line" with a half-width of an order of magnitude of 1 cm^{-1}, at room temperature. The intensity is fairly strong; so it is possible (and necessary) to record the spectrum with narrow slits. However, the slit function must be taken into account in the analysis of the experimental results obtained at room temperature. When several bands are superimposed, for instance in a set of "hot-bands", the Q-branches will often be resolved.

But, at this stage, we must show why the trace intensity spreads within a finite spectral range. With spectral terms of the form given by Eqs (10), (11), or (12), since none of the rotational quantum numbers change in a Raman transition giving rise to trace scattering, the rotational shifts for all these lines would be zero. Actually, the rotational constants A, B and C depend slightly on the vibrational state according to

$$B_{[v]} = B_e - \sum_i \alpha_i^B (v_i + \frac{d_i}{2}) \quad , \tag{17}$$

in which B_e is the rotational constant when all the nuclei are at equilibrium positions, α_i^B is a vibration-rotation interaction constant related to the ith vibrational mode and d_i the degeneracy of this mode. The subscript [v] designates the set of vibrational quantum numbers describing the vibrational state of the molecule. Similar expressions hold for A_v and C_v .

The rotational Raman shifts of the Q-branch lines of a linear molecule, in a $v_i = 1 \leftarrow v_i = 0$ transition, are

$$\Delta \tilde{\nu}_{rot} = B_1 J(J+1) - D J^2(J+1)^2 - [B_0 J(J+1) - D J^2(J+1)^2]$$

$$= - \alpha_i J(J+1) \qquad (J = 0,1,2....). \qquad (18)$$

The lines are actually very close as the α_i are generally very small (10^{-3} B order of magnitude).

The distribution of the intensity of the anisotropy scattering spreads over a large spectral range (about 50 cm^{-1} at room temperature). Since the intensity is small, one must use wide slits, but as the distribution is a smooth one, the deconvolution of the experimental contour by a slit function is not necessary. When a set of "hot bands" appears, the analysis becomes quite impossible. As the rotational shifts are multiples of the rotational constants A, B or C, the simplest expressions for the rotational terms are generally accurate enough to account for the experimental results. This type of study is of interest for the determination of molecular constants, such as the Coriolis interaction coefficient ζ. Analyses by computer simulation and analytical expressions of such distributions have been published.[8,9,10,12] A comparison between the two methods is given in Ref (9).

VII.   TRACE INTENSITY DISTRIBUTION

We shall now give some details about distribution studies of trace intensity by means of analytical expressions. For the linear molecules, the symmetric tops, and the spherical tops, general expressions are available. We shall give a detailed calculation for the linear molecule only.

A. Linear Molecules

The principle of the calculation can be found in Ref (11), but we give here a more detailed discussion.

The total Raman shifts, for a Q-branch, are given by

$$\Delta \tilde{\nu}_{Ram} = \Delta \tilde{\nu}_{vib} + \Delta \tilde{\nu}_{rot}$$

$$= \Delta \tilde{\nu}_{vib} - \alpha_i J(J+1) \qquad (19)$$

according to Eq (18), for a transition $1 \leftarrow 0$ involving the ith normal mode.

The absolute frequencies of the Raman lines in the Stokes spectrum are

$$\tilde{\nu}_{\text{Ram. Stokes}} = \tilde{\nu}_{\text{exc.}} - \Delta\tilde{\nu}_{\text{Ram}}$$

$$= \tilde{\nu}_{\text{exc.}} - \Delta\tilde{\nu}_{\text{vib}} + \alpha_i J(J+1) ,$$

in which $\tilde{\nu}_{\text{exc.}}$ is the frequency of the exciting line.

The frequency of the band origin is

$$\tilde{\nu}_{\text{vib}} = \tilde{\nu}_{\text{exc.}} - \Delta\tilde{\nu}_{\text{vib}}$$

and the Raman frequencies referred to the band origin are

$$\Delta\tilde{\nu} = \tilde{\nu}_{\text{Ram}} - \tilde{\nu}_{\text{vib}} = \alpha_i J(J+1) . \tag{20}$$

We emphasize here the meaning of our notation: the $\Delta\tilde{\nu}_{\text{Ram}}$ are the usual Raman shifts, while the $\Delta\tilde{\nu}$ are the frequencies referred to the band origin. In the Stokes spectrum, their variations are of opposite signs.

The intensities of the lines, for trace and anisotropy scattering (see Ref. 2), are

$$I_{\text{tr.}} (J) \propto C_{\text{tr.}} (2J+1) \exp [- B J(J+1) hc/kT] \tag{21}$$

$$I_{\text{anis.}} (J) \propto C_{\text{anis.}} (2J+1) \frac{J (J+1)}{(2J-1)(2J+3)} \exp [- B J(J+1) hc/kT] \tag{22}$$

Usually, $C_{\text{tr.}} \gg C_{\text{anis.}}$. Moreover, the matrix element $J(J+1)/(2J-1)(2J+3)$ approaches 1/4 with increasing J. Thus, we may deal with the total intensities in the case of the linear molecule, namely:

$$I (J) \propto (2J+1) \exp [- B J(J+1) hc/kT] . \tag{23}$$

At this stage, considering only experimental situations where Bhc << kT, it is possible to introduce simplifying approximations, such as

$$\Delta\tilde{\nu} \simeq \alpha_i J^2 \tag{24}$$

and

$$I(J) \propto 2J \exp\left[-Bhc\, J^2/kT\right]. \tag{25}$$

For the linear molecules, these approximations are <u>not</u> necessary, but as they are generally introduced in more cumbersome situations, we will develop the calculation in that form.

We may write:

$$I(\Delta\tilde{\nu}) = I(J)\left|\frac{dJ}{d(\Delta\tilde{\nu})}\right|, \tag{26}$$

where $\left|dJ/d(\Delta\tilde{\nu})\right|$ is the density of lines. The absolute value must be taken, as $dJ/d(\Delta\tilde{\nu})$ may be a negative quantity.

Since $\qquad\qquad J = (\Delta\tilde{\nu}/\alpha_i)^{1/2}$,

$$\left|dJ/d(\Delta\tilde{\nu})\right| = \frac{1}{2(\alpha_i\,\Delta\tilde{\nu})^{1/2}}, \tag{27}$$

and $\qquad\qquad I(J) \propto 2(\Delta\tilde{\nu}/\alpha_i)^{1/2} \exp\left[-Bhc\,\dfrac{\Delta\tilde{\nu}}{\alpha_i}/kT\right]. \tag{28}$

Using Eq (26), we obtain

$$I(\Delta\tilde{\nu}) \propto \frac{1}{|\alpha_i|} \exp\left[-a\,\Delta\tilde{\nu}\right], \tag{29}$$

in which

$$a = Bhc/\alpha_i\, kT.$$

It is also possible to use reduced coordinates; viz.:

$$I(x) \propto e^{-x}. \tag{30}$$

The reader should understand that the results we have obtained would be the same without any approximations.

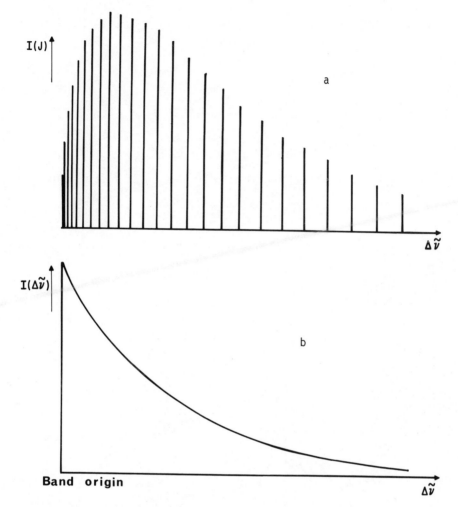

Figure 3. Q-branch of a linear molecule. a - Resolved b - Unresolved.

According to Eq (24), $\Delta\tilde{v}$ and α_i have the same sign, as have $\Delta\tilde{v}$ and a, therefore, in Eq (29). Thus, x may take only positive values in Eq (30).

In Figure 3, we see the conversion of a resolved contour into an unresolved one, for the Q-branch of a linear molecule, with emphasis on the action of the density of lines.

The convolution by a slit function, with the assumption that it has a gaussian shape, is easy in this case. If $\tilde{v}_{app}$ is the frequency given by the spectrometer, $\exp[-h(\tilde{\nu}_{app} - \tilde{\nu})^2]$ is the slit function, and

$$\Delta\tilde{\nu}_{app} = \tilde{\nu}_{app} - \tilde{\nu}_{vib},$$

then the convolution product gives

$$I(\Delta\tilde{\nu}_{app}) \propto \exp(-[a\,\Delta\tilde{\nu}_{app} - \frac{a^2}{4b}]) \cdot \text{erf c}\,[\frac{a}{2b^{1/2}} - b^{1/2}\Delta\tilde{\nu}_{app}]. \quad (31)$$

In Figure 4, the result of the convolution by a gaussian slit function is shown.

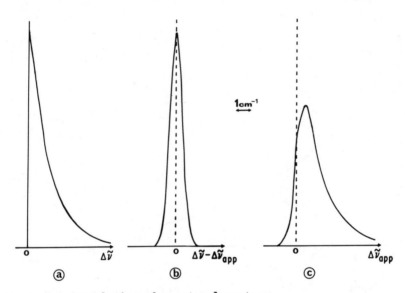

Figure 4.   Convolution of a natural contour.

a - Natural contour of a Q-branch of a linear molecule. Here $Bhc/\alpha kT = \ln 2$, so that $\Delta\tilde{\nu}_{1/2} = 1$ cm^{-1}.

b - Slit function. In order to keep a constant area for the distribution functions under the convolution, the gaussian function is normalized:
$$(b/\pi)^{1/2}\exp[-b(\Delta\tilde{\nu} - \Delta\tilde{\nu}_{app})^2],$$
and $b = 4\ln 2$, so that $\Delta\tilde{\nu}_{1/2} = 1$ cm^{-1}.

c - Convoluted contour.

B. Spherical Tops

The spectral lines, as given by Eq (12), are the same for linear molecules and spherical tops, so that $\Delta\tilde{v} \simeq \alpha_i J^2$ again. On the other hand, as the degeneracy is now $(2J+1)^2$ (Sec. III), the intensities for trace scattering, which is the only contribution, are

$$I(J) \propto (2J+1)^2 \exp\left[-B\,J(J+1)\,hc/kT\right] . \qquad (32)$$

The natural distribution is thus given by

$$I(\Delta\tilde{v}) \propto \frac{1}{2|\alpha_i|^{3/2}} (\Delta\tilde{v})^{1/2} \exp - [a\,\Delta\tilde{v}] , \qquad (33)$$

in which, once again,

$$a = \frac{Bhc}{\alpha_i\,kT} .$$

The result is the same as given in Eq (29) except for the term $(\Delta\tilde{v})^{1/2}$ arising from the different degeneracy.

With reduced coordinates, Eq (33) may be written as

$$I(x) \propto e^{-x} . \qquad (34)$$

The analytical convolution is also possible, and gives, using the same slit function as above:

$$I(\Delta\tilde{v}_{app}) \propto \exp\left[-b\,\Delta\tilde{v}_{app}\right] \int_0^\infty X^{1/2} \exp\left(-[X^2+A\,X]\right) dX , \qquad (35)$$

in which $A = b^{1/2}(a - 2b\,\Delta\tilde{v}_{app})$.

In Figure 5, a natural contour is shown.

C. Symmetric Tops

A detailed calculation may be found in Ref. (13).

As the spectral terms given in Eq (11) depend on two quantum numbers and contain two different rotational constants (B and C for the oblate top), the problem is somewhat more cumbersome than the previous

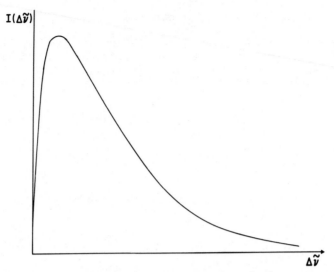

Figure 5.   Natural contour of a Q-branch of a spherical top.

ones.   We give only the main features of the calculation and the results.
Using the same notation as above,

$$\Delta \tilde{\nu} = \alpha^B \, J(J + 1) + (\alpha^C - \alpha^B) \, K^2. \tag{36}$$

For trace scattering, the partial intensities are given by

$$I(J,K) \propto 2 \, (2J+1) \, \exp \left\{ - [B \, J \, (J+1) + (C-B)K^2] \right\} hc/kT. \tag{37}$$

Using the notation

$$\gamma = \frac{C}{B} - 1, \qquad \gamma' = \frac{\alpha^C}{\alpha^B} - 1, \qquad \text{and} \qquad \sigma = \frac{Bhc}{kT} \, ,$$

and performing the usual approximations, we find

$$\Delta \tilde{\nu} = \alpha^B \, (J^2 + \gamma' \, K^2) \, , \tag{38}$$

and

$$I \, (J,K) \propto J \, \exp \, [-\sigma \, (J^2 + \gamma \, K^2)] \, . \tag{39}$$

From this, it is possible to work out an expression of the distribution in a sub-branch, that is, a series of lines corresponding to the same value of $|K|$, designated by K_i, J running from K_i to infinity. There is an infinity of such sub-branches and the intensity distribution in the branch results from an integration. According to the signs of $(\gamma - \gamma')$, α^B, and α^C, the results are somewhat different. In what follows, we assume that $\alpha^B > 0$.

1) $(\gamma - \gamma') > 0$

(a) $\alpha^C > 0$. Then

$$I (\Delta\tilde{\nu}) \propto \exp\ [-\sigma\, \Delta\tilde{\nu}/\alpha^B] \cdot \text{erf}\ [\frac{\sigma\,(\gamma - \gamma')}{\alpha^C}\, \Delta\tilde{\nu}]^{1/2}\ . \tag{40}$$

The branch spreads only in the $\Delta\tilde{\nu} > 0$ range.

(b) $\alpha^C < 0$. Then

$$I (\Delta\tilde{\nu}) \propto \exp\ [-\sigma\, \Delta\tilde{\nu}/\alpha^B\ ] \tag{41}$$

for the $\Delta\tilde{\nu} > 0$ range, and

$$I (\Delta\tilde{\nu}) \propto \exp\ [-\sigma\, \Delta\tilde{\nu}/\alpha^B\ ] \cdot \text{erf c}\ [\frac{\sigma\,(\gamma - \gamma')}{\alpha^C}\, \Delta\tilde{\nu}]^{1/2} \tag{42}$$

for the $\Delta\tilde{\nu} < 0$ range.

2) $\gamma - \gamma' < 0$

(a) $\alpha^C > 0$. Then

$$I (\Delta\tilde{\nu}) \propto \exp\ [-\sigma\, \Delta\tilde{\nu}/\alpha^B] \cdot \int_0^A \exp X^2\, d X\ , \tag{43}$$

in which

$$A = [\frac{\sigma\,(\gamma - \gamma')}{\alpha^C}\, \Delta\tilde{\nu}\ ]^{1/2}\ .$$

The branch spreads only in the $\Delta\tilde{\nu} > 0$ range.

(b) $\alpha^C < 0$. This possibility does not occur. Since

$$\gamma - \gamma' = \frac{C}{B} - \frac{\alpha^C}{\alpha^B}$$

and C/B is a positive quantity, when $\alpha^C < 0$ and $\alpha^B > 0$, $(\gamma - \gamma')$ cannot be negative.

If $\alpha^B < 0$, all the previous expressions hold for opposite signs of α^C and $\Delta\tilde{\nu}$. For instance, for $\alpha^B < 0$ and $\alpha^C > 0$, one must use Eq (41) for the negative range of $\Delta\tilde{\nu}$, and Eq (42) for the positive range.

More condensed expressions may be obtained by putting $\alpha^B C/\alpha^C B = \varepsilon$, and using the reduced coordinate

$$x = \sigma \Delta\tilde{\nu}/\alpha^B = \frac{Bhc}{\alpha^B kT} \Delta\tilde{\nu} .$$

Eq (40) may thus be written

$$I(x) \propto e^{-x} \cdot \text{erf} \, [(\varepsilon - 1) x]^{1/2} . \tag{44}$$

and so on.

In Figure 6, typical distributions for $\alpha^B > 0$ and $(\gamma - \gamma') > 0$ are shown.

D. Asymmetric Tops

No decisive work has been done on the Q-branches of the asymmetric top; we shall give only some research ideas. In the trace scattering spectrum, it seems that transitions between corresponding rotational levels only may occur, with an intensity proportional to $(2J + 1) \exp [-E_{rot}/kT]$. If the rotational levels are known up to the J-values for which the population becomes negligible in the experiment under consideration, and if the three interaction constants α^A, α^B, and α^C are available from infrared spectra, it should be possible to do a computer simulation of the branch. It seems out of scope to obtain rigorous analytical expressions.

VIII.   PRACTICAL DEVELOPMENTS

Up to this date, experimentation has concerned linear molecules essentially,[7,11,14] but the potentialities are the same for all types of rotators.

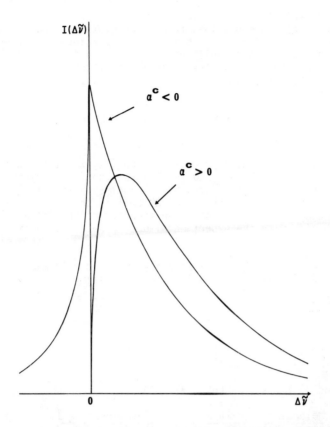

Figure 6. Natural contours of a Q-branch of a symmetrical top.
Here, $\gamma - \gamma'$ and α^B are assumed to be positive.

As the natural profile of a Q-branch depends on the parameter a of
Eq (29), that is $Bhc/\alpha_i kT$, its analysis allows a measurement of the
molecular parameter B/α_i, or a temperature measurement. The method
appears to be of interest for the determination of the quantities α_i,
as pointed out in Ref. (11) and (14). The temperature may also be
determined with the same accuracy. Considering the natural distribution,
for the sake of simplicity, we can see that the half bandwidth is
proportional to the absolute temperature:

$$T = \frac{Bhc}{\alpha_i k \ln 2} \Delta\tilde{\nu}_{1/2} .$$

(45)

The sensitivity of the method is determined by the numerical value
of B/α_i which may vary in a large range, depending on the anharmonicity
of the vibrator. An error less than 20°K seems reasonable in typical
cases, when α_i is known with sufficient accuracy.

Actually, the temperature obtained from a bandwidth is a rotational one, which may be different from the vibrational temperature when the system is out of thermodynamic   equilibrium. The vibrational temperature may be determined from the ratio of the intensities of the fundamental and of the "hot bands". This ratio is

$$\frac{I_{(v+1) \leftarrow v}}{I_{1 \leftarrow 0}} = (v+1) \exp [- v \omega hc/kT] . \qquad (46)$$

In a system at thermodynamic equilibrium, using Eq (31) for the contours of the Q-branches of all the bands, and Eq (46) for the intensity ratio of the different bands, it has been possible to analyse a system of thirteen bands lying in a spectral range of 20 cm^{-1};[13] this system is the v_1 band of S_2 C and 12 "hot bands". The accuracy of the fit of the calculated distribution with the experimental one is better than 1% of the full scale.

However, in some cases, Eq (46) apparently does not hold and further experimental and theoretical investigations seem necessary.

REFERENCES

1. G. Placzek and E. Teller, Z. Phys. 81, 209 (1933).

2. G. Herzberg, Molecular Spectra and Molecular Structure I.
    Spectra of Diatomic Molecules, 2nd Ed., (D. Van Nostrand Co., Inc.,
    Princeton, 1950).

3. G. Herzberg, Molecular Spectra and Molecular Structure II. Infrared
    and Raman Spectra of Polyatomic Molecules, (D. Van Nostrand Co.,
    Inc., Princeton, 1945).

4. B. P. Stoicheff, in Advances in Spectroscopy, Vol. 1, ed. by H. W.
    Thompson, (Interscience, New York, 1959), p. 91.

5. R. Gaufrès and S. Sportouch, J. Mol. Spectry. 39, 527 (1971).

6. J. Herranz and B. P. Stoicheff, J. Mol. Spectry. 10, 448 (1963).

7. M. Lapp, L. M. Goldman and C. M. Penney, Science 175, 1112 (1972).

8. F. N. Masri and N. H. Fletcher, J. Chem. Phys. 52, 5759, (1970).

9. S. Sportouch and R. Gaufrès, J. Chim. Phys. 69, 470 (1972).

10. S. Sportouch, C. Lacoste and R. Gaufrès, J. Mol. Struct. 9, 119 (1971).

11. R. Gaufrès and S. Sportouch, Compt. Rend. Acad. Sci. Paris <u>272</u>,
     995 (1972).

12. R. J. H. Clark and D. M. Rippon, J. Mol. Spectry. <u>44</u>, 479 (1972).

13. R. Gaufrès and S. Sportouch, in <u>Advances in Raman Spectroscopy</u>,
     Vol. 1, ed. by J. P. Mathieu, (Heyden and Son, Ltd., London, 1973),
     Chapt. 62.

14. S. Sportouch, Thèse d'Etat Montpellier 1972 (Dissertation).

GAUFRES DISCUSSION

BERSHADER - I would like to ask a question about the polarizability
you were talking about -- either the trace or the off diagonal polarizability components. What is their variation with vibrational quantum
number? Certainly, in the case of atomic species, the polarizability is
a strong function of the quantum number of the level. Now here, you
assume the polarizability is constant.

GAUFRES - Actually, when I write a transition moment for the vibrational Raman effect, the polarizability elements must be thought as
functions of the vibrational coordinates, according to

$$\alpha_{jk} = \alpha^0_{jk} + \sum_{i=1}^{3N-6} \left(\frac{\partial\alpha_{jk}}{\partial Q_i}\right)_o Q_i$$

say a constant part, and a function of all the normal coordinates. The
first term accounts for the Rayleigh scattering, and the second one for
the vibrational Raman scattering. The expansion of α_{jk} as a function of
the vibrational coordinates is the classical expression of the dependence
of the polarizability on the vibrational state.

HILL - I thought that in a molecule like water the Q-branch lines
are sufficiently spread out so that you don't see a grouping of them.

GAUFRES - In the Raman spectrum, as in the infrared, the Q branch of
an asymmetric top may spread on a large spectral range, due to the relaxation of the selection rules, which allows transitions between rotational
levels of the same J, (Q branch) but of different τ. Nevertheless, all
the transitions $J'_{\tau'} \leftarrow J''_{\tau''}$ with $J' = J''$ and $\tau' = \tau''$ give rise to a bunch
of lines of close frequencies. In the Raman spectrum, the whole trace
scattering is contained in these lines. By considering only trace scattering, one drops many frequencies of the spectrum, and the Q branch itself
is much simplified. Moreover, the intensity rules are very simple for
trace scattering. Generally, trace scattering is much stronger than anisotropy scattering, and one observes in a totally symmetric band of an
asymmetric top a strong sharp line, essentially due to trace scattering.

Then it is possible, as a first approximation, to omit anisotropy scattering in such a simplified situation. But you must experimentally verify that anisotropy scattering is very weak in comparison to trace
scattering.

LAPP - Are there any significant differences between homonuclear and
heteronuclear diatomic molecules for the calculation of Q-branch profiles?

GAUFRES - Not in the smoothed spectra. The intensity alternation
among lines originated from levels of even or odd J values averages out
when one considers the overall effect. Strictly speaking, the intensity
distribution in a Q branch of a homonuclear diatomic molecule is the sum
of two distributions, one for the even lines, one for the odd ones, which
are the same apart from the constant factor $(I + 1)/I$. For more complicated molecules with some symmetry there is also a nuclear spin degeneracy
factor, which depends on J, but as J increases, this term very quickly
approaches a constant. So, I think that, in every case, the nuclear spin
degeneracy may be omitted when one calculates smoothed spectra.

HENDRA - Don't you have to worry in the case of a linear molecule
about a possible change of symmetry in it's first excited vibrational
state? Doesn't this effect influence the question under discussion,
because the intensity rules for the rotational lines, that is J odd and
J even allowed or not allowed, breaks down if the symmetry of the first
excited vibrational state is different from that of the ground state?

GAUFRES - In what molecules?

HENDRA - The classic work in this respect was done by J. J. Barrett
and A. Weber [J. Opt. Soc. Am. 60, 70 (1970)] on CO_2. The rotation of
CO_2 in excited states of the vibrational bending mode does not obey the
same selection rules as the rotation of the molecule in the vibrational
ground state. In the high temperature measurements that we are concerned
with at this meeting, the population of the excited vibrational state at
667 cm-1 is quite high, and so you must allow for this. This is clearly
seen in the rotational Raman spectrum of CO_2. Does this affect the Q-
branch vibrational scattering as well? I think it does!

BARRETT - At room temperature, about 7% of the CO_2 molecules are on
the state at 667 cm^{-1}, and you do observe pure rotational scattering in
that state for all J-values. For the ground vibrational state, the antisymmetric rotational states (odd J-values) are not populated due to the
zero nuclear spin of the oxygen atoms.

HENDRA - But how does this phenomenon effect the Q-branch width?

GAUFRES - In the CS_2 (or CO_2) molecule, the Q-branch of the band
$(1, 1^1, 0) \leftarrow (0, 1^1, 0)$, for instance, contains twice as many lines as
that for the band $(1, 0°, 0) \leftarrow (0, 0°, 0)$. The question is the same as
that for the homonuclear molecules: The half-band width, and more generally the profile, is the same for both Q-branches, but the intensity of
the band arising from the two-fold degenerate excited level must be

multiplied by a factor 2. We actually analyzed the whole ν_1-region of CS_2 in this way, assigning the weight factor 2 to this type of hot band, and the fit between the observed contour and the calculated one was excellent. For the hot bands of the type $(v + 1, 0°, 0) \leftarrow (v, 0°, 0)$, we used the factor $(v + 1)$.

HARVEY - With respect to the spectral position of the Q-branches for a molecule like CO_2, I think the answer to the question is as follows. If there is strong coupling between two vibrations -- one perhaps low-lying and thermally well-populated -- then the Q-branch of the other can be affected. For example, hot bands arising from coupling which Herzberg calls $\nu_1 + \nu_2 - \nu_2$ should introduce additional Q-branches. [N.B. Editors: See pp. 266-269 of Ref. 3 of the paper under discussion.]

HENDRA - I'll have to think about that!

GAUFRES - There is another question that could be asked: What is the influence of pressure or temperature broadening of the individual lines on a profile? As long as the broadening of the individual lines is not too important, it has no influence on the intensity distribution in an unresolved profile. In the calculation of the analytical expression, I assumed that the distribution in an individual line is a δ function. But this assumption is not necessary, as long as the finite width of a rotational line remains negligible in comparison to the width of the overall profile.

MEASUREMENT OF AIRCRAFT TURBINE
ENGINE EXHAUST EMISSIONS

by

Donald A. Leonard

Avco Everett Research Laboratory, Inc.
Everett, Massachusetts 02149

ABSTRACT

This paper describes a current program to demonstrate the use of laser Raman backscattering as a means to measure the composition and temperature of aircraft engine exhaust emissions. The physical characteristics of turbine exhausts that most significantly influence Raman spectroscopic analysis are the relatively low pollutant concentrations in combination with elevated temperatures such that significant upper level excitations can occur with subsequent broadening and overlapping of the Raman spectra. We discuss the effects of these characteristics regarding quantitative gas analysis by Raman scattering at elevated temperatures. An experimental lidar unit was constructed using a 0.5 watt average power 3371 Å pulsed nitrogen laser, a 24-inch diameter Dall-Kirkham Cassegrainian-type receiver telescope and a 1 meter, double scanning spectrometer for spectral analysis. The entire system operation is controlled by computer, including calibration, data collection and data analysis. Preliminary experimental results will be presented.

I. INTRODUCTION

A program is currently underway at the Avco Everett Research Laboratory, Inc. to demonstrate the use of laser Raman scattering as a means to measure the composition and temperature of aircraft engine exhaust emissions.[1,*] Hardware has now been assembled in a portable unit and simulated exhaust stream tests will soon be conducted in cooperation with Avco Lycoming.

---

*This work is being supported by the Air Force Aero Propulsion Laboratory.

The physical characteristics of turbine exhaust emissions that most
significantly influence Raman spectroscopic analysis are the relatively
low pollutant concentrations (~ 100 PPM) and elevated temperatures
(~ 1000°K). The elevated temperatures produce significant upper level
excitations with consequent broadening and overlapping of the Raman spectra. These effects must be taken into account if Raman scattering is to
be used for quantitative gas analysis at elevated temperatures.

II. THEORETICAL CONSIDERATIONS

The photon energies observed in Raman scattering differ from the
photon energy of the incident light by an energy corresponding to some
difference in allowed energy levels of the scattering molecule. The
molecular energy shift can be either positive or negative, i.e., energy
can be either subtracted from ("Stokes process") or added to ("anti-
Stokes process") the incident photon. The turbine exhaust measurement
system uses "Stokes" processes since they provide the best indication of
concentration when the ground states are most highly populated.

Photons of energy $h\nu_0$ are incident on the molecule, which is originally in rotational level J_m of vibrational level v_k. Photons of energy
$h\nu_{RAMAN}$ are scattered, and the molecule is left in the higher vibrational
energy state v_e, with rotational quantum number J_n. The difference in
energy between the incident and scattered photons is exactly equal to the
energy added to the molecule, i.e.,

$$h\nu_o - h\nu_{RAMAN} = E(v_e, J_n) - E(v_k, J_m)$$

The rules governing which transitions among the levels in a molecule
are allowed for Raman scattering are called "selection rules." The selection rules for a diatomic molecule are:

$$\Delta v = 0, \pm 1$$

$$\Delta J = 0, \pm 2$$

The same rules generally apply separately to each vibrational mode in
polyatomic molecules, although special circumstances may act to modify the
rules, e.g., "Fermi resonance" effects such as in CO_2.

At low temperatures all the lines having $\Delta J = 0$ lie very close to
each other in energy and are not resolved except with very high resolution
spectroscopy. The $\Delta J = \pm 2$ transitions, however, are well separated in
energy and appear as side bands on either side of the intense $\Delta J = 0$ line.

Figure 1 is a schematic representation of the vibrational Raman spectra of
turbine exhaust emissions at typical concentrations and at room temperature (300°K). The wavelength scale is appropriate for excitation with a
3371 Å pulsed nitrogen laser. It is to be noted that the relative intensity of the Raman scattering is plotted as a logarithmic scale.

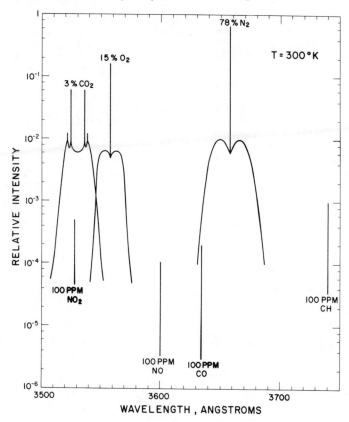

Figure 1. Schematic of Vibrational Raman Spectra of
Typical Turbine Exhaust Emissions at 300°K.

As seen in this figure the dominant signal is from the nitrogen which
appears at about the same strength (78% ± 1%) over typical fuel/ air ratios
and can be used with corresponding precision as an internal self-calibra-
tion and normalizing standard. That is, all the other specie signals are
normalized to the nitrogen signal. Thus the measurement is not affected
by variations in laser power, optical transmission losses (i.e., smoky
exhausts), or drifts in electronic detection efficiencies.

The pollutants of interest are shown on this figure at the 100 PPM
concentration level. As can be seen: the nitric oxide (NO) and the

hydrocarbon (CH) signals appear as strong signals without interference; the carbon monoxide (CO) signal appears as somewhat less strong than the nitrogen $\Delta J = -2$ side band; and the nitrogen dioxide (NO_2) signal is approximately 2 orders of magnitude less than the overlapping side bands of the carbon dioxide (CO_2) doublet and is seriously masked by it.

As the temperature of the turbine exhaust gas under consideration is increased, the Raman spectra becomes more complex, the rotational $\Delta J = \pm 2$ bands broaden, more $\Delta J = 0$ lines appear, and all $\Delta J = 0$ lines broaden into bands with slightly shifted peak values. These effects must be correctly taken into account if Raman scattering is to be used for quantitative analysis at elevated temperatures.

The physical basis for these effects is twofold: Firstly, as the temperature increases the populations of the various excited levels increase exponentially with temperature according to the appropriate Boltzman factor and degeneracy for the level in question. Secondly, because of the anharmonicity of molecular vibrational potentials and the vibration-rotation interaction the energy levels of the excited states are not uniformly spaced, as would be those of an isolated harmonic oscillator and rigid rotor.

The energy of a diatomic molecule in vibrational state (v) and rotational state (J) can be written in the following way (following the notation of Herzberg):

$$E\ (v,J) = \omega_e\ \left(v + \frac{1}{2}\right) - \omega_e x_e\ \left(v + \frac{1}{2}\right)^2 + \omega_e y_e\ \left(v + \frac{1}{2}\right)^3 + \ldots \qquad \text{(VIB.)}$$

$$+ B_e J\ (J + 1) - D_e J^2\ (J + 1)^2 + \ldots \qquad \text{(ROT.)}$$

$$- \alpha_e\ \left(v + \frac{1}{2}\right) J\ (J + 1) - \beta_e\ \left(v + \frac{1}{2}\right) J^2\ (J + 1)^2 + \ldots \qquad \text{(VIB-ROT.)}$$

As indicated, this represents the energy of level (v,J) as composed of three components, the vibrational energy written as an expansion in powers of (v + 1/2), the rotational energy written as an expansion in powers of J (J + 1), and the vibration-rotation interaction energy written as an expansion in powers of J (J + 1) with (v + 1/2) as a coefficient. For purposes of turbine exhaust analysis over the temperature range of interest (i.e., 900-1200°K) calculations have shown the terms with coefficients ω_e, $\omega_e x_e$, B_e and α_e to be significant. Keeping only these significant terms, we find that the energy absorbed by a diatomic molecule for $\Delta v = 1$ and $\Delta J = 0$ is

$$E\ (v + 1,J) - E\ (v,J) = \omega_e\ -2\ (v + 1)\ \omega_e x_e\ -\alpha_e J\ (J + 1).$$

The two negative terms in this expression produce a smaller energy separation between the laser line and the Raman lines, thus producing a shift toward shorter wavelengths (a spectral "blue shift") of the Raman spectrum. These negative terms become increasingly important as the temperature is raised and levels corresponding to higher values of v and J become increasingly populated.

Again keeping only the significant terms, the energy change for $\Delta v = 1$ and $\Delta J = \pm 2$ is

$$E(v + 1, J \pm 2) - E(v,J) = \omega_e - 2(v + 1)\,\omega_e x_e \pm 4\,B_e\left(J + \frac{1}{2} \pm 1\right).$$

The above expressions together with the selection rules determine the spectral positions of the various possible vibrational Raman transitions. The strength of a transition is given by the following expression:

$$S(v,J) \propto \sigma\,(v + 1)\,N\,(v,J)$$

where σ is the vibrational Raman cross section.

Figure 2 is a schematic representation of the vibrational spectrum at a temperature of 1000°K of the same turbine exhaust composition which was shown in Figure 1 at a temperature of 300°K. The effects predicted by the preceding theoretical considerations are clearly visible in this figure.

Measurement at the 100 PPM level of both NO and CH still appears to be straightforward and without interference. The situation with respect to CO has become worse at the higher temperature since the rotational sideband of N_2 has broadened and increased in intensity at the wavelength which corresponds to the spectral position of the CO line. At 1000°K the N_2 sideband is about 60 times greater than the signal from 100 PPM of CO.

The interference of the $\Delta J = \pm 2$ sidebands with respect to the $\Delta J = 0$ signals can be reduced by means of polarization-sensitive detection. Figure 3 is a plot which shows at a temperature of 1200°K a calculation of the N_2 Raman band at 3658 Å and the CO Raman band at 3634 Å at a CO concentration of 100 PPM relative to the nitrogen.

As shown in Figure 3 the effect of the interference is reduced by the use of polarized light. A factor of two reduction has been both calculated and measured for plane polarization and a factor of six has also been both calculated and measured experimentally for circular polarization. With circular polarization the ratio of the contribution of the CO Raman scattering at 3634 Å to the sum of the CO and N_2 Raman scattering at 3634 Å is about 10% for a 100 PPM mixture of CO in N_2. Thus CO should be measurable at the 100 PPM level.

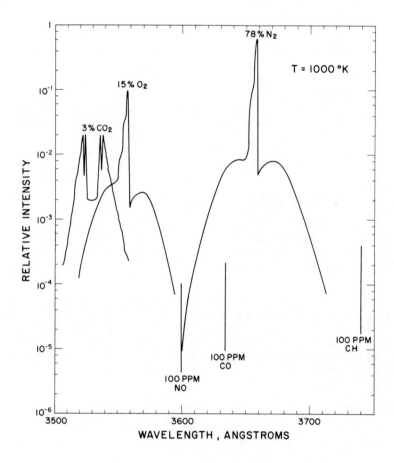

Figure 2.   Schematic of Vibrational Raman Spectra of
             Typical Turbine Exhaust Emissions at 1000°K.

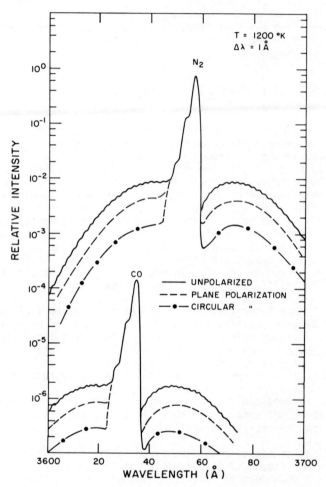

Figure 3.   Computed Raman Intensity for Nitrogen at 1200°K
and for Carbon Monoxide at 100 PPM Relative to N_2.

Another important change in the character of the turbine exhaust Raman spectra as the temperature is increased is the appearance of two secondary peaks in the CO_2 doublet which at 1000°K become as strong as the original room temperature doublet. These new lines correspond to transitions which originate from the lowest excited vibrational level of CO_2, the 010 level.

The 010 level is appreciably populated even at room temperature. Measurement of the ratio of the secondary to the primary peak has been found to be a sensitive means of temperature measurement throughout the temperature range of interest for turbine engine exhausts.

III. HARDWARE CONSIDERATIONS

In considering the use of a laser Raman system for gas turbine engine emissions analysis, many different operating situations may be encountered. These could range all the way from the analysis of the exhaust of a small engine on a fixed test stand to the remote tracking and analysis of the emissions of a large C5A during routine taxi, take-off and landing procedures at an operational airfield. The system therefore should be single ended with the laser transmitter and the Raman receiver contained on the same rigid, steerable and mobile platform. A rectroreflector or focusing element was also considered for placement directly behind the exhaust for increased signal collection but was ruled out of our conceptual design on the grounds of (1) severe vibration/alignment problems in the high level acoustic environments and (2) incompatibility with single-ended remote tracking capability as a system growth possibility.

As seen in Figure 4 the system is built as two transportable units. The transceiver package contains the laser, transmitter and receiver optics, monochromator and detection subsystems and supporting equipment such as cooling, pumping and power conditioning. The operator enclosure package contains the control and display panel and an enclosure for the operator and is capable of remote positioning from the systems package and turbine exhaust region for operator safety and comfort. A dark background plate of anodized aluminum, placed in the field of view of the receiver, behind the exhaust, is required to reduce daylight background interference to negligible levels.

The optical schematic of the transceiver system is shown in Figure 5. The basic design philosophy is to optically match the entrance slit of the monochromator to a mode control spatial filter slit whose image is coherently amplified within the laser. The laser output is then imaged by the transmitter mirror onto the exhaust and the scattering from the exhaust is then re-imaged by the collector mirror onto the entrance slit of the monochromator with near 100% efficiency. The slits can be oriented to sample a maximum number of streamlines in the flow for averaging purposes, or a minimum number of streamlines if concentration and temperature profiles are desired. The overlap region between the fields of view of the transmitter and receiver is controlled by the angle between their axes and the width of the spatial filter slit image.

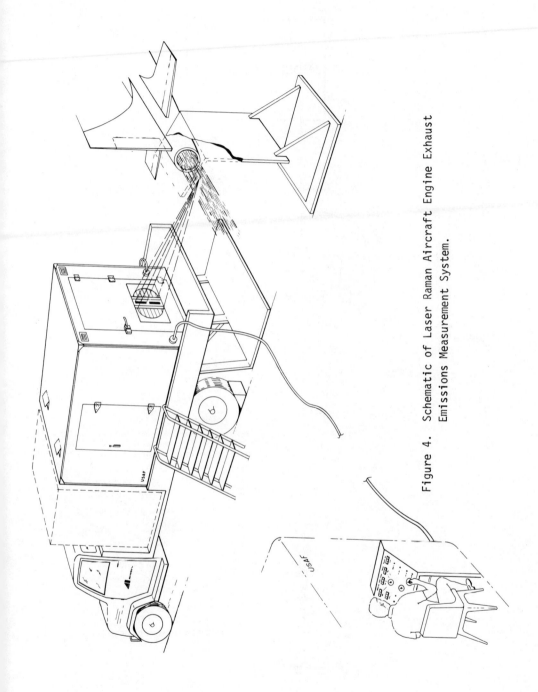

Figure 4.   Schematic of Laser Raman Aircraft Engine Exhaust Emissions Measurement System.

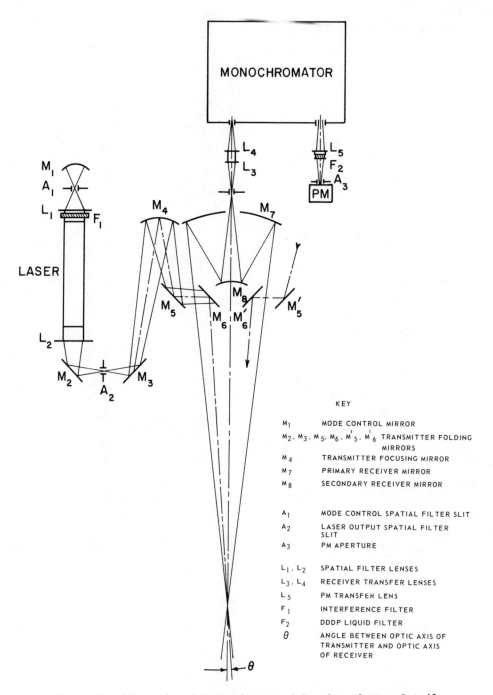

Figure 5.   Schematic of Transmitter and Receiver System Details.

The choice of laser wavelength depends on a variety of factors, among which are: (1) the availability of high power lasers in each wavelength region, (2) the $(1/\lambda)^4$ dependence of the Raman scattering cross section, which favors the shortest possible wavelength (except for occasional possible enhancements by resonance Raman effects), (3) photoelectric efficiencies, which are highest below 5000 Å, (4) solar background, which falls off rapidly below 3000 Å (solar blind region), (5) unfavorable absorptions, which indicate wavelength regions for certain situations must be avoided, e.g., atmospheric absorption below 2500 Å and SO_2 absorption between 2500 Å and 3200 Å, (6) unfavorable fluorescences in certain wavelength regions, such as the very strong fluorescence from NO_2 at wavelengths longer than 3975 Å, and (7) eye safety problems, which are less serious below 3400 Å where the vitreous fluid is much less transparent with subsequently greatly reduced hazard to the retina.

Consideration of all these factors lead to the choice of the 3371 Å ultraviolet pulsed nitrogen laser which operates on the (0,0) transition of the second positive band system of the nitrogen molecule. The high pulse repetition rate and low energy per pulse of this laser also favor eye safety.

The monochromator in a laser Raman spectroscopic system usually has as its primary function the suppression of the direct non-frequency-shifted laser light that is strongly scattered by particulate matter or other solid objects in the field of view of the receiver optics. This is especially important in the Raman analysis of turbine exhausts, which contain significant quantities of particulate matter. Insertion of an isotropic bulk volume absorption filter in an appropriate position in the receiver optical train relaxes somewhat the monochromator requirements.

With the 3371 Å laser as a source, a good choice for such a filter is a water solution of 2, 7-dimethyl-3, 6-diazacyclohepta-1, 6-diene perchlorate (DDDP) which has the property of essentially complete absorption at 3371 Å, but with nearly complete transparency in the vibrational Raman region at wavelengths of 3500 Å and longer.

The electronics for the system are shown schematically in Figure 6. The entire system is controlled by a small minicomputer which receives instructions from the operator via a teletype and then proceeds to scan the monochromator, take in data, process the data and display results after suitable calculations have been performed. All calibrations are computer-controlled.

At low light levels the most efficient photoelectric detection method is actual quantum counting. However, this system had to be built to handle a relatively large dynamic range during each wavelength scan through the entire spectrum. To allow for many photoelectrons being produced in one laser pulse gate time, a charge to pulse train converter is used to digitize the photomultiplier signals. A digitized signal is thus taken into the computer at a 500 H_z rate. The ambient light level is also sampled at a 500 H_z rate between laser pulses and used as a continuous background correction.

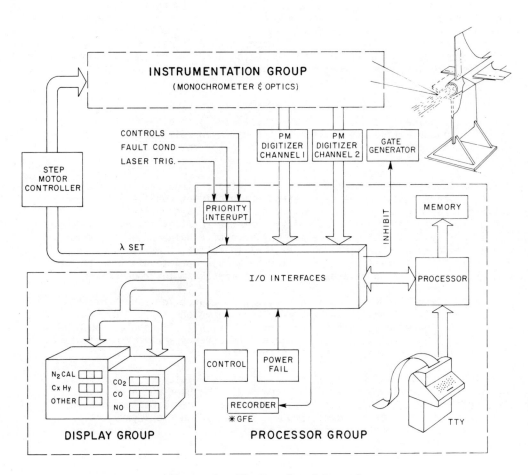

Figure 6.   Electronic   Schematic.

Calibration of the entire system is accomplished by measuring the Raman scattering signal from molecular nitrogen. The signals received from the Raman scattering from the various molecular species under analysis is ratioed to the signal received from the molecular nitrogen.

IV. PRELIMINARY DATA

The hardware described above has been assembled in a mobile van. The initial tests have involved observation of the Raman spectra of normal atmospheric air. The sample volume of air from which the Raman scattering was obtained was positioned 2 meters from the van and had approximate dimensions of 1 mm x 10 mm x 50 mm.

Figures 7, 8 and 9 show portions of the spectra obtained in the spectral regions corresponding to nitrogen, nitric oxide and water vapor. As shown, a dynamic range of about four orders of magnitude is available between the signal from N_2 in Figure 7 and the noise level at the NO

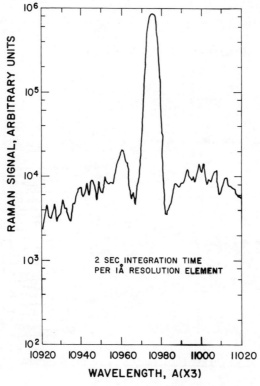

Figure 7. Raman Spectrum of Atmospheric Nitrogen.

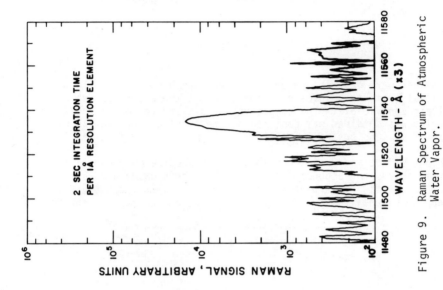

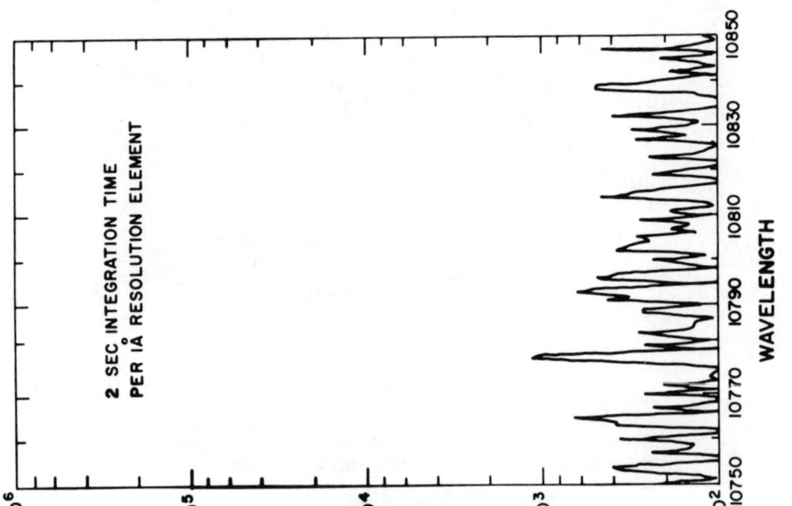

Figure 8. Noise Level at Spectral Position of NO Raman Line.

Figure 9. Raman Spectrum of Atmospheric Water Vapor.

spectral position as shown in Figure 8. Concentrations of NO at the
100 PPM level could thus be observable with this instrumentation. Shown
in Figure 9 is the scattering from atmospheric water vapor at a S/N of
about 100 and at a relative concentration of about 1% of the atmospheric
nitrogen signal in Figure 7. It should be noted that the wavelength
scale on these figures is 3 times the actual wavelength of the spectra.
(The grating was used in the 3rd order when this data was taken).

REFERENCES

1. D. A. Leonard, Avco Everett Research Laboratory Research Note 914,
    (1972).

    [N.B. Editors. See also G. E. Bresowar and D. A. Leonard, "Measure-
    ment of Gas Turbine Exhaust Pollutants by Raman Spectroscopy,"
    AIAA Paper No. 73-1276 (1973).]

LEONARD DISCUSSION

HEISER - How does the concentration value of 100 PPM for gases
such as NO stack up with what we have to measure in order to be able to
determine whether we are polluting or not? Is that big or small?

BRESOWAR - Well, 100 PPM NO would be a low estimate for high pressure
ratio engines, like the JT9D and the CF6. Since EPA standards will apply
to future engines, and the industry and DOD are leaning toward higher
pressure ratios, 100 PPM NO is a useful measurement range, perhaps 75%
less than state-of-the-art high pressure ratio engine capabilities at
takeoff or climbout. However, low pressure ratio aircraft gas turbines
generally will not produce 100 PPM NO except perhaps at very high power
settings. Our best forecasts indicate that considerable numbers of
these turbines will be produced after 1980, when regulations are to be
enforced, and hence a measurement technique that can measure 10-100 PPM
NO is needed. By the way, I still feel that Raman can do the job with
more efficient signal processing and hardware, which will be forthcoming
as the state of the art advances.

MACK - Do you think that low concentrations of CO can be measured
in air? Specifically, do you have a solution to the problem of interference
of the N_2 signature with that from CO?

LEONARD - I think that at 100 PPM CO, the N_2 side band will be about a
factor of 10 above the CO level with polarization-sensitive detection, so
really what we try to do first is to measure the profile of the N_2 side band
at that position and then subtract out the difference. In that kind of
processing, you can usually pick up a factor of about 10. Beyond that,
we have to start to think about using a narrow line laser. The laser we
are using now is the pulsed N_2 laser, which puts out an envelope of 3 or 4
rotational lines in a 1Å wide band. The separation of the N_2 side band
lines are also about 1Å. However the CO Q-branch ($\Delta J=0$) main peak, which
we are trying to look at, will be broadened at high temperature because of

the slightly shifted contributions of the higher J-levels. Thus,
even if we used a narrow-line laser, trying to look at CO through the
N2-side band lines is not very straightforward, and there are still
some remaining problems. If all of these problems are taken together,
I believe we can still measure CO at 100 PPM. Beyond that, further
improvements with narrow-line lasers are certainly possible.

PENNEY - Don, I'm impressed that you can have a photomultiplier
looking through a double monochrometer at a daylight scene.

LEONARD - I forgot to say it was after dark! Or, we have a dark
target in the field of view of the receiver.

PENNEY - Can you gate the photomultiplier so that you can operate
it in the daytime, or are there other procedures you can use?

LEONARD - We have made a calculation of the scattered sunlight
off the particles in the exhaust. Now we are talking about clean
exhausts, but they still have some particles in them. If you look at
the sunlight that scatters in the same wavelength band and in the same
time gate, you have a signal which is equivalent to a signal at the
10 to 100 PPM level. The results depend upon the character of the
particles in the exhaust. In my figure showing the truck, with the
laser on it, and the aircraft engine (Fig. 4), a screen was placed in
back of the exhaust. If you do get in trouble with sunlight, you put
the screen there.

GOULARD - If you wish to eliminate Rayleigh scattering due to sunlight
can't you do so by taking the intensity of the signal on both sides of those
Q branches and subtracting the average from the signal at the Raman
frequency?

LEONARD - We might gain another factor of 10 by subtracting the
background. We are looking at background for a long period of time in
between laser pulses, so we have a very accurate measure of it. But
there might still be some trouble, since you have to worry about statistical fluctuations in the large background signal.

BERSHADER - Would you comment with regard to the interaction
between the measured system and the probe? I am concerned about the
possibility that the incident light beam may significantly perturb the
gas even as you are trying to measure its properties. Is it possible to
show, for example, that the dipole interaction energy is small compared
to the level spacing, or that the energy absorbed by the gas is small
compared to its static enthalpy?

LEONARD - Well, the second question is the easiest one, because I
forgot the first one already! In our case, the laser average power is a
half watt, and each pulse is a millijoule. Also, the fraction of this
energy that is absorbed by the gas in the measurement volume is extremely
small. Thus there seems to be little likelihood that the gas will be
perturbed, considering the high power levels of a typical aircraft turbine

engine.   Could I have the first question again?

BERSHADER - The first question involved a comparison of the
dipole interaction energy with the spacing of the energy levels.   In
other words, does the focused laser beam significantly perturb the
energy levels?   The dipole interaction energy between the gas molecule
and the electric field of the light wave in ratio to any other
characteristic energy in the molecule tells you whether you are
perturbing anything in your system by the measurement technique.

LEONARD - Well you have to remember that scattering is an
effectively instantaneous process, so that very little energy is
built up in the gas during its course.   The fraction of molecules that
interact with the beam is extremely small, so I don't believe the
perturbation is significant.

BERSHADER - In any case where the molecular dipole is interacting
with the electric radiation field, the implicit assumption is made
that the interaction is linear, i.e., the polarizability is constant.
Otherwise, the quantum levels are disturbed and the problem becomes
nonlinear.   An "advance stage" of this problem is indicated by self-
focussing or thermal blooming of laser beams or actual breakdown of
the gas.   But this problem may begin at an earlier stage, when the
energy of the radiation field is comparable to molecular energies.

MACK - One effect of this type is the optical Stark effect, but
that doesn't become significant until much higher power densities than
those involved in Don Leonard's apparatus.

PENNEY - As Dr. Bershader has anticipated, it is well demonstrated
that a low power laser beam can perturb an absorbing gas (N. B. Editors,
see the paper by W. Kiefer in this Proceedings), and also that a light
pulse that is energetic, short and tightly focussed can significantly
perturb even a non-absorbing gas.   In fact, the onset of significant
perturbations seems to depend strongly on the type and state of the
observed gas, and as such, this onset determines a fundamental trade-off
between measurement accuracy, temporal resolution and spatial resolution
in a Raman scattering measurement.   Perturbations can often be reduced by
reducing the amount of integrated light energy in the incident beam
during the measurement time, by holding the energy constant but increasing
the measurement time, or by defocussing the beam.   But each of these steps
reduces one of the measurement qualities in the trade-off.   Fortunately,
in most of the gas probe applications we have discussed -- non-absorbing
gases at low temperature and weakly absorbing gases at high temperature --
the trade-off appears to be favorable; that is, it is well beyond the
measurement requirements.   Nevertheless, I believe that the questions we
have raised here are most appropriate, because they will become increasingly important as the types of application and limits of Raman scattering
probes are extended.

THE USE OF A FABRY-PEROT INTERFEROMETER FOR STUDYING
ROTATIONAL RAMAN SPECTRA OF GASES

by

J. J. Barrett

Materials Research Center, Allied Chemical Corporation
Morristown, New Jersey

ABSTRACT

A method is described that allows simultaneous observation of a
large number of rotational Raman lines of a gas. This method can
provide substantial advantages in light economy over techniques using
vibrational Raman scattering for the measurement of gas properties.
It employs a Fabry-Perot etalon whose free spectral range is adjusted
to coincide with the spacing between rotational Raman scattering lines.
Principles of operation, construction details of several instruments,
and experimental results for several gases are described.

I.   INTRODUCTION

A Fabry-Perot interferometer[1] has been used to study rotational
Raman scattering in gases. By not using any prefiltering of the
spectrum to separate overlapping orders, the free spectral range
of the interferometer can be adjusted so that the interference fringes
which are produced by the individual rotational lines of the spectrum
overlap to produce a single fringe whose intensity is essentially equal
to the sum of the intensities of the individual rotational lines.
The use of this interferometric technique offers three possible
advantages in light economy over the use of vibrational Raman scattering
for measurements of gas properties. The first advantage is that the
pure rotational scattering cross section in a molecule is usually
larger than the vibrational scattering cross section. Secondly, this
interferometric technique utilizes all the rotational Raman lines as a
single signal, such that the detected signal is substantially equal
to the sum of the signals from the individual rotational lines.
Recently,[2] experimental results have appeared in the literature concerning
the ratio of the Raman cross section of all the pure rotational lines
to that of the Q-branch of the ν_1 vibration in the same molecular

species. For example, this ratio for CO_2 was measured to be 175. The third possible advantage of this interferometric technique is that the luminosity of the Fabry-Perot interferometer is greater than that of the grating spectrometer for the same resolution. Jacquinot[3,4] has analyzed this problem in some detail and has shown that the gain in luminosity of the Fabry-Perot interferometer over the grating spectrometer (for a typical case) can be about two orders of magnitude. Therefore the Fabry-Perot technique utilizing rotational scattering can be at least three orders of magnitude more sensitive than the conventional grating spectrometer utilizing vibrational scattering.

II. MOLECULAR ROTATIONAL STRUCTURE AND ROTATIONAL RAMAN LINES

Before I discuss this interferometric technique, I would like to review the basic equations for a simple rigid rotator and the experimental method for obtaining the Raman spectrum of a gas. According to classical mechanics, the rotational energy of a rigid body is given by the equation

$$E = \frac{1}{2} I\omega^2 = \frac{P^2}{2I} , \qquad (1)$$

where I, ω, and P are the moment of inertia, angular frequency and angular momentum of the rigid body respectively. In terms of the quantum mechanical formulation, the angular momentum is quantized and is equal to $\sqrt{J(J+1)}\hbar$ where J is the rotational quantum number. The quantum mechanical energy of the rotating body is

$$E(J) = \frac{\hbar^2}{2I} J(J + 1). \qquad (2)$$

If we express equation (2) in units of wavenumbers (cm^{-1}), then we have

$$F(J) = \frac{E(J)}{hc} = BJ(J + 1). \qquad (3)$$

The quantity $F(J)$ is referred to as the term value of the rotational level J and the symbol B is the rotational constant defined as

$$B = \frac{h}{8\pi^2 cI} . \qquad (4)$$

For pure rotational S-branch scattering, the selection rule governing
the change in the rotational quantum number J is $\Delta J = 2$. The magnitude
frequency shift of the J-th rotational Raman line from the exciting
frequency (Rayleigh line) is

$$|\Delta \nu| = F(J + 2) - F(J) = 4B(J + 3/2).$$ (5)

The Raman lines which are shifted down in frequency with respect to
the Rayleigh line are called Stokes lines and those which are shifted up
in frequency are referred to as anti-Stokes lines. A schematic
representation of a rotational Raman spectrum is shown in Figure 1. The

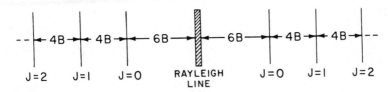

Figure 1.   Schematic representation of a pure rotational Raman
spectrum for the rigid rotator class of molecules. The
selection rule $\Delta J = +2$ gives rise to the S-branch
rotational lines which are symmetrically displaced on
both sides of the Rayleigh line corresponding to
Stokes and anti-Stokes frequency shifts.

Rayleigh line which is due to elastic scattering (the incident and
scattered photons have the same energy) is usually several orders of
magnitude more intense than the rotational Raman lines which are
produced by inelastic scattering. From equation (5), we see that the
frequency shift for the first rotational line is 6B and that the
frequency difference between adjacent rotational lines is equal to 4B.
This fact will be important in the discussion of the interferometric
technique.

III.   EXPERIMENTAL APPARATUS

The conventional apparatus for studying Raman scattering in gases
usually employs a laser and a grating spectrometer. A typical

experimental arrangement[5] is shown in Figure 2. The excitation source
was an argon ion laser and the gas sample was positioned at a focus
inside the laser cavity defined by the end mirrors M_1 and M_3. The
collection lens for the scattered light L_2 had a focal length of 50 mm
and a f/d ratio of 0.95. The Dove prism P_2 was used to rotate the line
image of the source so that it was aligned with the slit of the
spectrometer. The grating spectrometer had a focal length of 580 mm
and a 84 x 84 mm grating which was ruled at 1440 lines/mm. The light
at the exit slit of the spectrometer was detected using a cooled
EMI 6256S-A photomultiplier tube with a photon counting detection
system. The binary-coded-digital output of the counter was recorded
on magnetic tape and a computer was used to smooth digitally the
spectrum according to the procedure described by Savitzky and Golay.[6]
Figure 3 shows a pure rotational Raman spectrum of CO_2 as plotted
by the computer. The pressure of the sample of CO_2 was 40 Torr.
A series of weak lines are present which occur halfway between the
strong pure rotational ground state lines. These lines are due[7] to
pure rotational scattering from the 01^10 vibrational state which lies
at 667 cm^{-1} above the ground state. At 300°K, the 01^10 state has a
population of about 7 percent. This spectrum represents the first
observation of pure rotational Raman scattering in the 01^10 vibrational
state of CO_2.

    In order to study rotational Raman scattering from molecules which
are heavier than CO_2, higher resolving power than can be achieved by
using a moderate size grating spectrometer is necessary. Because of
this limitation in resolution, the use of a Fabry-Perot interferometer
for studying Raman scattering in gases was considered. The Fabry-Perot
interferometer consists of two flat mirrored plates which are aligned
parallel to each other. Interference fringes are produced by multiple
reflections of light between the mirrored surfaces. If I_i is the
intensity of the incident light then the intensity of the light (I_t)
which is transmitted by the Fabry-Perot interferometer is given by
the Airy function

$$I_t = \frac{T^2}{1 + R^2 - 2R\cos\delta} \cdot I_i \; , \tag{6}$$

where $T + R + A = 1$ and δ is the phase difference between the interfering
rays and is equal to

$$\delta = 4\pi\mu\omega d \tag{7}$$

for incident rays normal to the interferometer mirrors. The transmittance,
reflectance and absorptance of the Fabry-Perot mirrors are represented
by the symbols T, R and A respectively, μ denotes the refractive index
of the medium between the Fabry-Perot mirrors, d is the mirror
separation and the wavenumber ω (in units of cm^{-1}) is equal to the

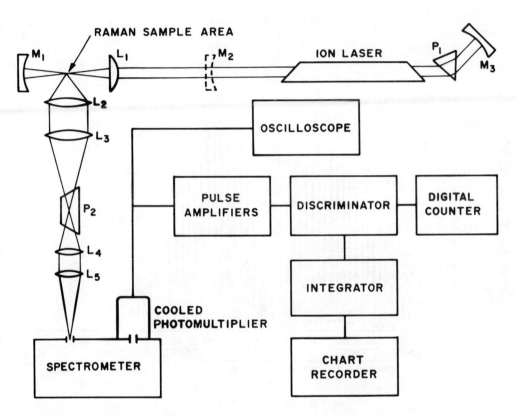

RAMAN SAMPLE AREA

M₁   L₁   M₂   ION LASER   P₁   M₃

L₂
L₃

P₂

L₄
L₅

OSCILLOSCOPE

PULSE
AMPLIFIERS

DISCRIMINATOR

DIGITAL
COUNTER

INTEGRATOR

COOLED
PHOTOMULTIPLIER

SPECTROMETER

CHART
RECORDER

Figure 2.   Experimental apparatus for obtaining the Raman spectrum of a
gas.   The excitation source is an ion laser and a grating
spectrometer is used to analyze the light which is scattered
by the gas.

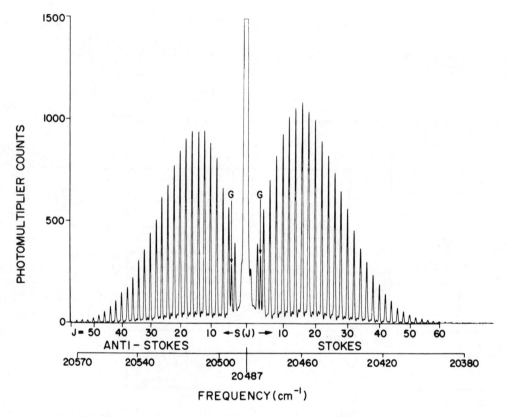

Figure 3.   The pure rotational Raman spectrum of CO_2 at a pressure of
40 Torr.   The 4880 Å line of an argon ion laser was used
as the excitation source.

reciprocal of the wavelength of the incident light. Transmission
maxima of I_t occur for $\cos\delta$ equal to unity. Hence

$$\delta = 2\pi m , \tag{8}$$

where m = 0, 1, 2, . . . and denotes the order of interference. For a
fixed value of the mirror separation d, I_t maxima occur for the
frequency interval $\Delta\omega$ of the incident light equal to

$$\Delta\omega = \frac{1}{2\mu d} , \tag{9}$$

where $\Delta\omega$ is known as the free spectral range of the interferometer.
Therefore the Fabry-Perot interferometer behaves like a comb filter
with transmission windows which are regularly spaced in frequency by
$\Delta\omega$. If the incident light has only one frequency component at ω,
then I_t maxima occur at mirror displacement intervals Δd equal to

$$\Delta d = \frac{1}{2\mu\omega} = \frac{\lambda}{2\mu} , \tag{10}$$

i.e., for a mirror displacement interval of a half wavelength.

If we illuminate a Fabry-Perot interferometer with monochromatic
light, we obtain an interference pattern which consists of a series
of concentric rings as shown in Figure 4. The light source which
produced this interference pattern was the 4880 Å line of an argon
ion laser oscillating in a single longitudinal mode. In order to
record photoelectrically the interference fringes, the fringe pattern
is imaged on a plane which contains a small circular hole or aperture
at the center of the circular fringe pattern. This aperture allows
only the light from the center fringe to be transmitted to a photomultiplier on the other side of the aperture. The interferometer is
scanned by changing the optical path between the interferometer mirrors.
If, for example, the interferometer is scanned by moving the mirrors
together a distance equal to a half wavelength, the center fringe
will disappear and the first ring will collapse to form the center
fringe. If the bright center fringe of Figure 4 corresponds to the
n-th interference order at 4880 Å, then the first interference ring
is the (n + 1)-th order at 4880 Å. However the center fringe could be
formed by the (n + 1)-th order for a wavelength slightly different than
4880 Å. In fact there is a whole series of wavelengths which differ
by the free spectral range and which will produce the center fringe.
This filtering property of the Fabry-Perot interferometer can be
utilized advantageously to transmit simultaneously all the rotational
Raman lines from a gas. The concept is schematically depicted in
Figure 5. A typical rotational Raman spectrum is represented
in Figure 5 (a) and in Figure 5 (b) the vertical lines represent
the transmission peaks of the Fabry-Perot interferometer. The

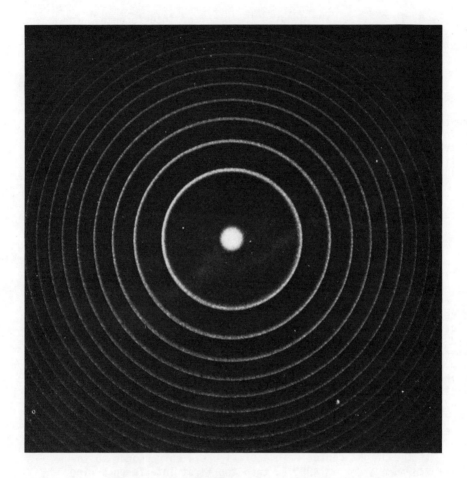

Figure 4.   Photograph of the Fabry-Perot interference fringes
            for 4880 Å illumination from an argon ion laser.

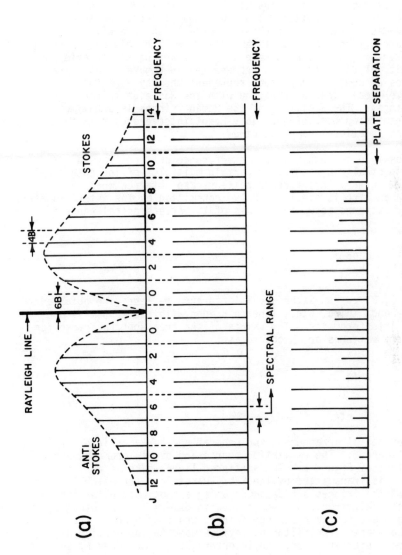

Figure 5. Schematic representation of (a) a rotational Raman spectrum, (b) the Fabry-Perot transmission comb and (c) the interferogram which is produced by scanning the interferometer in the region for which the free spectral range is equal to the rotational line separation.

frequency spacing between adjacent rotational Raman lines is essentially equal to 4B. By adjusting the free spectral range of the Fabry-Perot interferometer to be equal to 4B, it is then possible to achieve one-to-one correspondence between the transmission peaks of the Fabry-Perot interferometer and the rotational Raman lines from the gas. For this case all the rotational Raman lines will be transmitted as a single fringe and the Rayleigh scattered light will be blocked. As the interferometer is scanned a half order, the Rayleigh line will be transmitted and all the rotational Raman lines will be blocked. The resulting interferogram which is generated by scanning the interferometer in the vicinity of $d = 1/(8\mu B)$, is represented in Figure 5 (c). The vertical lines of constant amplitude represent the Rayleigh fringes and the smaller vertical lines alternating with the Rayleigh fringes are the Raman fringes. The amplitude of the Raman fringes is a maximum when the transmission comb of the Fabry-Perot interferometer is best matched to the rotational Raman spectrum.

It is possible to obtain other interference patterns for values of the free spectral range equal to certain multiples of the rotational gas constant. Primary interference patterns are obtained when every rotational Raman line is simultaneously transmitted by the interferometer. These primary patterns occur for values of the free spectral range $\Delta\omega$ equal to

$$\Delta\omega = \frac{4B}{2k - 1} \, , \tag{11}$$

where $k = 1, 2, 3, \ldots$. Secondary interference patterns are obtained when every second, third, etc., line coincides with the transmission peaks of the interferometer. The values of the free spectral range for which secondary patterns are produced is

$$\Delta\omega = 2k'B \, , \tag{12}$$

where $k' = 3, 4, 5,$ etc.

The experimental arrangement for generating the Raman interferograms is shown in Figure 6. The excitation source and sample geometry are the same as that of Figure 2. The scattered light which is collected by the lens L_2 is transmitted by the Fabry-Perot interferometer (mirrors M_4 and M_5) and the fringes are detected using a photomultiplier tube. A pulse counting electronic system is used to detect the photomultiplier signal and the detected interference fringes are displayed on the chart recorder. There are no prefiltering devices used in conjunction with the Fabry-Perot interferometer. An interferogram is produced by scanning one of the Fabry-Perot mirrors at a constant rate and recording the variations in the fringe intensity on the chart recorder.

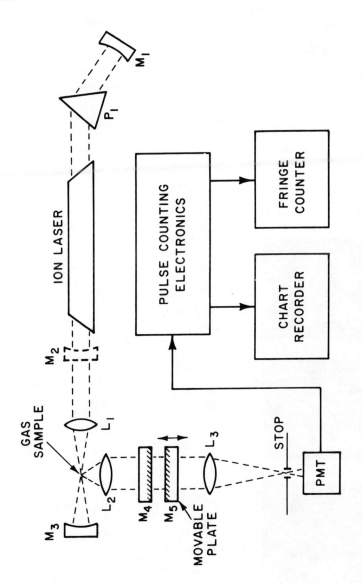

Figure 6.   The experimental arrangement which is used for the interferometric studies. The mirrors M_4 and M_5 represent the Fabry-Perot plates and the interferograms are produced by the uniform displacement of the mirror M_5 as indicated in the figure.

IV.   EXPERIMENTAL RESULTS

Some experimental results for N_2O gas at atmospheric pressure are shown in Figure 7. The ground state B-value for nitrous oxide is equal to 0.4190 cm^{-1} and hence the rotational line separation is approximately equal to 1.6 cm^{-1}. Figure 7 shows oscilloscope traces for the 8B and 6B secondary patterns and the 4B primary pattern. The most intense fringes are due to Rayleigh scattered light. In Figure 7 (a) there are two Raman fringes between adjacent Rayleigh fringes because for

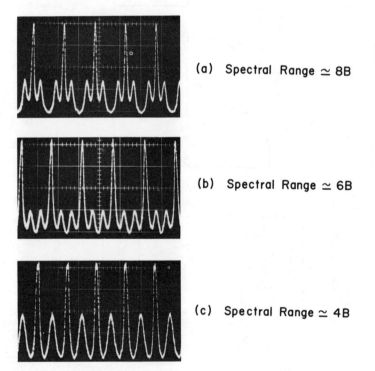

(a)   Spectral Range $\simeq$ 8B

(b)   Spectral Range $\simeq$ 6B

(c)   Spectral Range $\simeq$ 4B

Figure 7.   Oscilloscope traces of the Raman interferograms for N_2O.

this case, the Fabry-Perot interferometer transmits one set of Raman lines corresponding to every second rotational Raman line and then the other set of rotational Raman lines. For the 6B pattern shown in Figure 7 (b) the interferometer transmits the rotational Raman lines in three sets with every third Raman line being transmitted simultaneously in the same set. For this case, however, one set of Raman lines coincides with the Rayleigh fringe. In Figure 7 (c), the primary interference pattern is shown for which there is a one-to-one correspondence between the rotational Raman lines and the interferometer transmission windows. The rotational Raman fringes are most intense in Figure 7 (c) because for this case, essentially all the rotational Raman lines are transmitted to form a single fringe.

In order to permit continuous mechanical scanning of the Fabry-Perot interferometer, a special instrument[8] was designed and constructed. This instrument makes use of air bearings in its construction and it is schematically represented in Figure 8. Two cylindrical air bearings are used to support a hollow stainless steel cylinder which has the moving interferometer mirror attached to one end. The stationary interferometer mirror is mounted in a holder which provides the necessary angular adjustments required for aligning the interferometer. A precision screw is used to provide the linear motion for scanning the interferometer. With this interferometer it was possible to scan through several thousand orders with no observable degradation in the alignment.

By using this air bearing interferometer, the complete 4B interferogram for N_2O gas was obtained and it is shown in Figure 9. The fringes of approximately constant intensity are due to Rayleigh scattering at 4880 Å. The rotational Raman fringes are located halfway between the Rayleigh fringes. The intensities of the Raman fringes in the vicinity of the maximum of the Raman-fringe profile are quire large, permitting the Raman radiation to be observed visually.

Figure 10 shows the complete 8B interferogram for N_2O. This secondary interferogram is produced when the interferometer is scanned in the region for which the free spectral range is equal to 8B. The appearance of two Raman fringes between adjacent Rayleigh fringes is due to the transmission of alternating sets of rotational Raman lines.

An improved version of the air bearing Fabry-Perot interferometer has been constructed recently and Figure 11 is a photograph which shows a closeup view of the interferometer mirrors and the moving carriage. The moving carriage is supported by three air bearings which ride on three parallel Invar rods. When the air bearings are pressurized, the carriage floats and does not make contact with the Invar supporting rods. The use of three air bearings in this configuration insures a purely linear motion of the carriage to a high degree of precision, thereby eliminating misalignment problems during a scan. The stationary mirror is mounted in a gimbal which is provided with piezoelectric micrometer screws for precision alignment control. The moving carriage is driven by making contact with a 50 cm long bar.

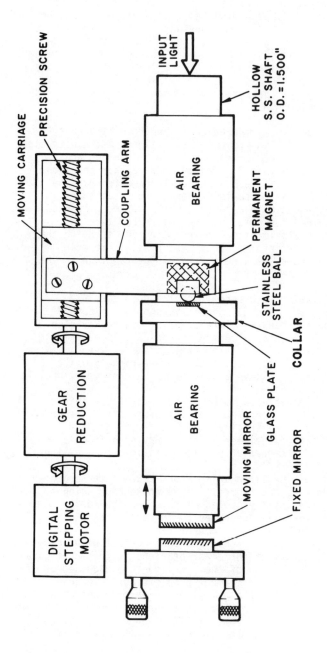

TOP VIEW

Figure 8.   A schematic diagram of the air bearing Fabry-Perot interferometer.   The light
to be analyzed enters the interferometer through the hollow stainless steel
cylinder which is supported by the air bearings.

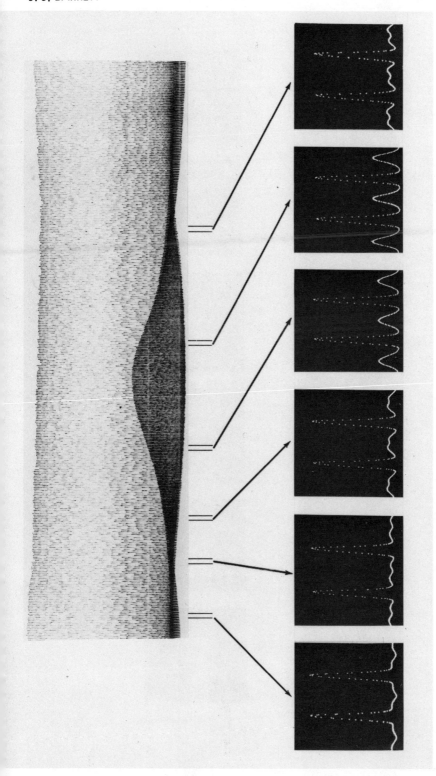

Figure 9. The complete 4B interferogram for N_2O. The lower part of the figure shows oscilloscope traces of various parts of the interferogram. The dotted appearance of the fringes in the oscilloscope traces is due to the digital sampling method which was used to record the interferogram.

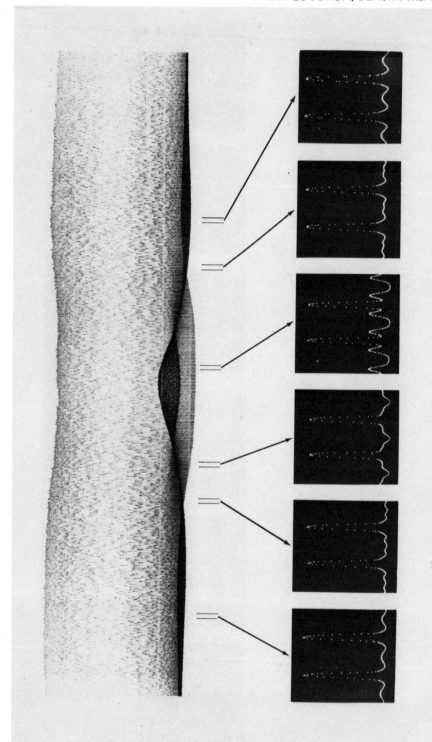

Figure 10.   The complete 8B interferogram for N_2O.

Figure 11.   A closeup side view of the improved version of the air bearing interferometer. Most of the interferometer is constructed of Invar.

One end of the bar is mounted in a bearing such that it is free to
pivot about a vertical axis. The other end of the bar is driven by a
high precision screw. Provision is made to be able to adjust the
position of the moving carriage contact point with respect to the
pivot point of the bar. This arrangement makes it possible to demagnify
the linear motion of the precision screw at the moving carriage
thereby minimizing screw errors at the moving carriage. An overall
view of the interferometer is shown in Figure 12. The massive
structure on the right is the basic interferometer which contains the
fixed and moving mirrors. It is constructed of Invar exclusively in
order to minimize dimensional changes due to ambient temperature
variations. The precision screw and the driving motor are shown just
to the left of the center of the photograph.

      The 4B interferogram of CO_2 gas at a pressure of one atmosphere
was generated using this improved interferometer and it is shown in
Figure 13. During the course of the experiments it was discovered
that the polarized Rayleigh scattered light could be significantly
reduced relative to the depolarized Raman scattered light by the use
of a polarizing element in the scattered light path. For this arrangement the rotational Raman fringes at the center of the 4B maximum
are about a factor of two more intense than the Rayleigh fringes.

      The possibility of using this Fabry-Perot technique for the
quantitative detection of pollutant gases in the atmosphere is currently
being investigated. A preliminary experiment involved the detection of
ambient CO_2 at a concentration of about 300 ppm in air. The 8B region
was investigated since it was found to be relatively free of background
due to Raman scattering from oxygen and nitrogen. The experimental
results are shown in Figure 14. The 8B interferogram for pure CO_2
at 760 Torr pressure is shown at the left in Figure 14. The fringe
labeled R is the Rayleigh fringe and the fringes labeled A and B
are the Raman fringes. The interferometer was piezoelectrically
scanned about a half order on each side of the Rayleigh fringe. The
gas cell was filled with known amounts of CO_2 gas in air and the Raman
fringes A and B were recorded as the concentration of the CO_2 was
decreased. At the extreme right in Figure 14, the interferogram of
ordinary room air is shown and the Raman fringes A and B due to CO_2
are clearly visible. Also present in this interferogram are two other
Raman fringes which are believed to be due to Raman scattering by
atmospheric nitrogen. The peaks of the Rayleigh fringes were
electronically reduced by a factor of ten in order to keep them on the
chart recording.

      Several experimental techniques are now being considered for
eliminating the background problem due to oxygen and nitrogen. These
experimental results will be reported in the near future.

Figure 12. An overall view of the interferometer showing the drive mechanism for scanning the interferometer. A solid granite optical table with vibration isolation legs is used to support the interferometer, laser and other optical elements used in the experiment.

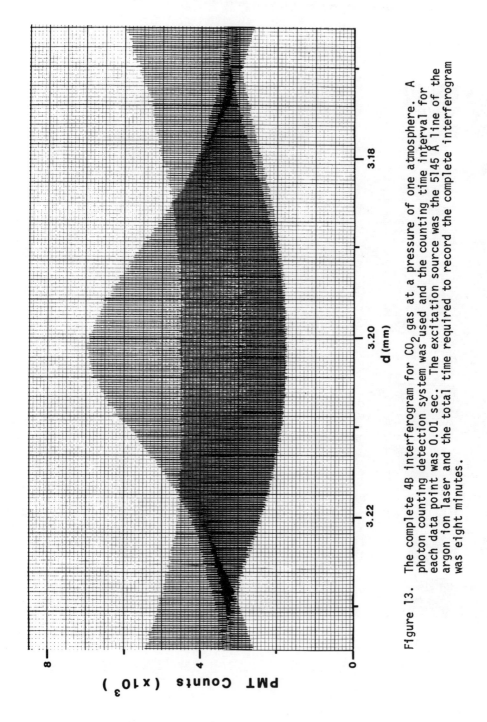

Figure 13.   The complete 4B interferogram for CO_2 gas at a pressure of one atmosphere. A photon counting detection system was used and the counting time interval for each data point was 0.01 sec. The excitation source was the 5145 Å line of the argon ion laser and the total time required to record the complete interferogram was eight minutes.

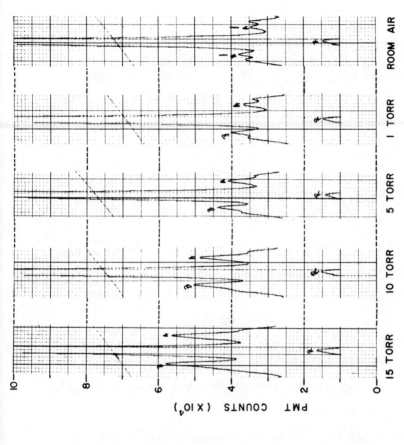

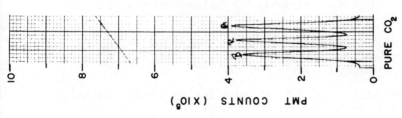

Figure 14. The 8B interferograms for pure CO_2 and for mixtures of known amounts of CO_2 in air. At the extreme right, the interferogram for room air is shown and the Raman fringes (A and B) corresponding to CO_2 at a concentration of about 300 ppm are clearly visible. The counting time interval was one second and the interferometer was scanned at a rate of one order in 2.8 minutes. The 5145 Å line was used as the excitation source.

NOTE ADDED IN PROOF

See the note added to the presentation of A. B. Harvey in this Proceedings concerning the application of the interferometric "comb" technique to the measurement of gas temperature.

REFERENCES

1.   J. J. Barrett and S. A. Myers, J. Opt. Soc. Am. 61, 1246 (1971).

2.   W. R. Fenner, H. A. Hyatt, J. M. Kellam and S. P. S. Porto, J. Opt. Soc. Am. 63, 73 (1973).

3.   P. Jacquinot, J. Opt. Soc. Am. 44, 761 (1954).

4.   P. Jacquinot, Rep. Prog. Phys. 23, 267 (1960).

5.   J. J. Barrett and N. I. Adams III, J. Opt. Soc. Am. 58, 311 (1968).

6.   A. Savitzky and M. J. E. Golay, Anal. Chem. 36, 1627 (1964).

7.   J. J. Barrett and A. Weber, J. Opt. Soc. Am. 60, 70 (1970).

8.   J. J. Barrett and G. N. Steinberg, Appl. Optics 11, 2100 (1972).

BARRETT DISCUSSION

LEONARD - Could I ask what the finesse of your interferometer was?

BARRETT - The finesse of the interferometer was measured to be 38.

LEONARD - How is the finesse defined?

BARRETT - It is the ratio of the free spectral range to the instrumental halfwidth, which was experimentally measured using a single frequency argon laser.

MACK - In terms of practical application, mechanical etalon holders are generally very good only for short periods of time. Would your etalon hold its alignment for periods of days?

BARRETT - The etalon that I built, with the one mirror floating on air, will stay in alignment for several days without trouble. Also, it is possible to scan it very fast by hand, over thousands of orders, without any misalignment of the mirrors. It is really quite good in this respect.

SALZMAN - I was wondering, what other molecules do you think you could detect?

BARRETT - Well, one of the next things I am going to look at is CO. The problem with CO is that it has a B-value very close to that of N_2, but I think I have a way of getting around that problem so that the fringes due to CO and N_2 will be separated sufficiently.

SALZMAN - I see. But it would seem that a number of molecules will have overlapping line positions in the rotational spectrum.

BARRETT - The main interference comes from atmospheric oxygen and nitrogen, and I think that I have a good way to alleviate that problem.

GAS CONCENTRATION MEASUREMENT BY COHERENT RAMAN ANTI-STOKES SCATTERING

by

P. R. Régnier and J.-P. E. Taran

Office National d'Etudes et de Recherches Aérospatiales (ONERA)
92320 Châtillon, France

Presented by J.-P. E. Taran

ABSTRACT

A novel technique for gas concentration measurements is described. This technique makes use of coherent Raman anti-Stokes scattering, which can be made much stronger than ordinary Raman scattering under conditions that are practical for gas probing applications. The intensity is sufficient to allow an image of the spatial distribution of a minor constituent (H_2 at about 100 to 1000 ppm) to be recorded on photographic film. In this presentation, we describe the basic phenomena, potential measurement capabilities, and the results of feasibility experiments.

I. INTRODUCTION

Spontaneous Raman scattering has turned into a powerful tool for pollution monitoring[1] and gas concentration measurement in aerodynamic flows[2]. Unfortunately, the useful signal levels are low; therefore, statistical noise and background light poses a serious problem in both types of measurements; furthermore, the determination of the concentration distribution of a gas across an aerodynamic flow (e.g. in flow mixing or in a combustion) is time consuming and cannot be done on a single laser shot.

The purpose of this paper is to show that stimulated Raman scattering (SRS) is an attractive alternative because the scattered light levels are many orders of magnitude higher, both in the Stokes and in the anti-Stokes sidebands. Unfortunately, stimulated scattering into the Stokes sideband will not occur unless the gas partial pressure is very large; besides, the Stokes intensity is so critically dependent on the laser pump intensity and the gain in the medium under study[3] that the readings would not be very meaningful. (Measuring the Raman gain as in Ref. 4

directly gives the gas concentration, which is proportional to the gain;
however, the method would not be accurate enough for partial pressures
below a fraction of an atmosphere.) But there is one process associated
with SRS that does not present this drawback. This process, called three-
or four-wave mixing by several authors[5-7], is held responsible for the
generation of the anti-Stokes sidebands in SRS. It reflects the coupling
of the laser pump and the Stokes wave through the resonant medium; for
that reason (see below) we preferred to call it coherent Raman anti-
Stokes scattering (CRAS).

Historically, this process was used first by Maker and Terhune[5] for
resonant susceptibility measurements in liquids, by Rado[6] and Mayer and
coworkers[8] for nonresonant susceptibility measurements in gases, and by
De Martini et al.[9] for H_2 line profile determination. Here we consider
application of this process to the determination of gas concentrations
and flow visualization. An experimental investigation on H_2 gas has
also been conducted in order to verify its feasibility.

II. THEORY

The interaction is best treated classically. Assuming collinear
plane waves with frequencies ω_L and ω_S propagating in the +z direction
in a homogeneous gas mixture (Figure 1) we consider Raman active

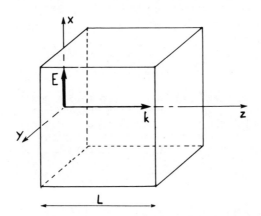

Figure 1. Field and wave vectors in the gas sample.

molecules that scatter with a frequency shift ω_v that is near resonance with $\omega_L-\omega_s$; i.e., $\omega_v \simeq \omega_L-\omega_s$. We can show that a wave at $\omega_a = \omega_L+\omega_v = 2 \omega_L-\omega_s$ is generated with an intensity[10,11]:

$$I_a = (\frac{4 \pi^2 \omega_a}{c^2})^2 I_L^2 I_s \left| \int_0^L \chi e^{i\Delta kz} dz \right|^2 \tag{1}$$

where $\Delta k = 2 k_L-k_s-k_a$ is the momentum mismatch, I_L and I_s are the laser and Stokes intensities respectively (which are assumed undepleted by the interaction), L the distance travelled into the medium and χ the complex Raman susceptibility:

$$\chi = N \frac{2 c^4}{\hbar \omega_s^4} \frac{d\sigma}{d\Omega} \frac{1}{2\Delta\omega-i\gamma} + \chi^{NR} \tag{2}$$

In Eq. (2), N is the number density of the resonant molecules (assumed to be in their ground state); $d\sigma/d\Omega$ is the differential Raman scattering cross section per molecule; γ is the damping constant; $\Delta\omega = \omega_v-(\omega_l-\omega_s)$; and χ^{NR} is the real, nonresonant contribution to the Raman susceptibility from all the molecules in the mixture.

A question arises as to the exact nature of the generation of the anti-Stokes wave described by Eq. (1). Two different processes that can generate these waves have been described by Bloembergen[12]:

(1) A parametric process associated with the real part of χ; two laser quanta at ω_L scatter into a quantum at ω_s and one at ω_a (Figure 2a) or vice versa, depending on the relative phases of the waves ($e^{i\Delta kz}$ term in Eq. (1)).

(2) A Raman process associated with the imaginary part of χ; two simultaneous transitions between the same initial and the same final states are involved (Figure 2b). The net result is an absorption at ω_s and a generation at ω_a, or vice versa, also depending on the phases of the waves.

The anti-Stokes generation, which is due to a polarization component of third order in the field strength, has been listed as a three[5] or four[6] wave mixing process. We can be somewhat more specific by calling it coherent Raman anti-Stokes scattering[13], or CRAS, which reflects the fact that the anti-Stokes wave is scattered by molecules with coherently driven vibration. This is the terminology we prefer. Note that the expression "parametric four wave mixing" used by Barak and Yatsiv[7] and by Sorokin et al.[14] for conversion in metal vapors is appropriate only for real susceptibilities and should not be used here.

CRAS will permit the determination of the number density of a molecular gas, provided all the parameters in Eqs. (1) and (2) are known. Let us now review the properties of these parameters.

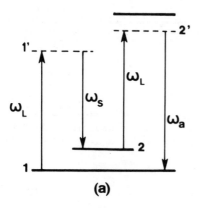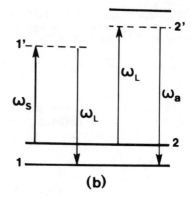

Figure 2.   Coherent Raman anti-Stokes scattering. Level 1 is the ground state; level 2 is any Raman-active level. (In the experiments described, 2 is a vibrational level.) The scattering process is instantaneous. Resonance enhancement can take place if either of the virtual levels 1' and 2' is close to an absorption. a) parametric process corresponding to the real part of the susceptibility - the transition cycle leaves the molecule in its ground state; b) Raman process associated with the imaginary part of the susceptibility - the medium gains or loses two quanta for each transition cycle. (This quantum picture should not be taken seriously, as discussed in Ref. 12.)

A. Phase Mismatch

The exponential in Eq. 1 can be taken as unity provided $z < l_c$, where l_c is given by $\Delta k l_c = \pi$. The mismatch Δk for parallel propagation vectors is a function of the medium dispersion and of the frequency difference $\omega_L - \omega_S \simeq \omega_v$. One has[8]

$$l_c = (\pi c / \omega_v^2) \left[2 \frac{\partial n}{\partial \omega} + \omega_L \frac{\partial^2 n}{\partial \omega^2} \right]^{-1} ,$$

where n is the index of refraction. For air under standard temperature and pressure conditions, l_c is given in Table 1. We can see that l_c is in general comparable to or larger than the thicknesses of most samples or wind tunnel flows that have to be analyzed.

Table 1

Frequency shifts of some gases, with corresponding spontaneous Raman
scattering cross sections and coherence lengths for four-wave mixing
in air.

gas	$\omega_v(cm^{-1})$	$(\frac{d\sigma}{d\Omega})_{gas} / (\frac{d\sigma}{d\Omega})_{N_2}$ [†]	$l_c(cm)$
O_3	1103	4.0	325
SO_2	1151	5.5	260
N_2O	1287	2.7	210
CO_2 {	1286	1.0	210
	1388	1.5	180
O_2	1556	1.2	140
NO	1877	0.55	98
CO	2143	1.2	75
N_2	2331	1.0	63
H_2S	2611	6.6	51
CH_4 {	2914	8.0	38.5
	3020	0.79	38
H_2O	3652		26
H_2	4155	2.2	20

$$(\frac{d\sigma}{d\Omega})_{N_2} = (4.4 \pm 1.7) \ 10^{-31} \ cm^2/sr$$

[†]After Chang and Fouche[1]

B.   Spectral Properties of χ

Far from an electronic absorption, χ does not depend strongly
on laser frequency. Thus going to shorter wavelengths does not
present the same advantage as with spontaneous Raman scattering
(with the ω^4 dependence of the scattering cross section). But in
the vicinity of an electronic absorption χ may gain several orders
of magnitude, with a resultant increase in scattered intensity I_a.
The variations of χ versus $\Delta\omega$, as given by Eq. (2), are plotted in
Figure 3. χ' and χ'' are the real and imaginary parts of the first,
resonant term; χ'' has the same spectral dependence as the spontaneous Raman line.

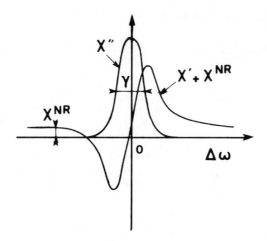

Figure 3.   Profile of the anti-Stokes Raman susceptibility for
a pure gas in the vicinity of a vibrational line.
The quantity χ^{NR} is not drawn to scale, being 4 to 5
orders of magnitude smaller than χ'' at $\Delta\omega = 0$.

In spontaneous Raman scattering experiments, the number of scattered
photons is independent of the linewidth γ of the Raman line. This is not
the case here, at least at resonance ($\Delta\omega = 0$), and γ is an important
parameter of the interaction. Unfortunately, although the Raman frequencies and   cross sections are well known, little has been published so far
on their linewidths, except for H_2[15,16] and N_2[15]. The case of H_2 is
quite peculiar in that only a few distinct Q-branch lines are observed at
room temperature, owing to the large rotational constant B; furthermore,
these lines have small widths ($\gamma/2\pi c \simeq 3\times10^{-2}$ cm^{-1})[15,16] and large separations. For other gases, the Q-branch lines cannot be resolved in
general, resulting in one broad line the width of which is determined by
the variations of B with vibrational quantum number and by the temperature. Furthermore, foreign gases, together with temperature and pressure,
will influence the linewidth γ and the vibrational frequency ω_v.

C. Detuning

The quantity $\Delta\omega$ is also an important parameter, which can be adjusted
at will, depending on the application considered. For $\Delta\omega = 0$, the
sensitivity is maximum. This sensitivity is determined by the ability to

distinguish the contribution of the resonant part of the susceptibility of the molecules to be detected from the "background light", which is produced by the nonresonant susceptibility of the buffer gas. But since both χ^{NR} and $d\sigma/d\Omega$ are practically independent of the nature of the gas, at least within factors of 2 or 3, the minimum concentration that can be detected in a gas mixture is determined primarily by the linewidth. For H_2, the minimum detectable concentration is thus of the order of 10 ppm, assuming that the buffer gas causes little line broadening. For other gases, this figure lies in the 10 to 100 ppm range.

As mentioned above, ω_v depends to some extent on temperature, pressure and foreign gas. For experimental situations where appreciable variations of these parameters are encountered, one may prefer to detune the excitation as was suggested in Ref. 8. For $\Delta\omega \gg \gamma$, the response of the medium is independent of γ and of the small shifts of ω_v, which are generally comparable to γ. The measurement then depends only on detuning, which can be set or determined quite accurately; but one then suffers a loss in the detectivity.

D. Beam Geometry

Two classes of problems are encountered experimentally:

a) visualization of a gas distribution across a whole aerodynamic field (in a wind tunnel, for instance), and in a fashion similar to interferometric or Schlieren methods;

b) point gas concentration measurements.

For the first type of application, parallel beams are to be used, both for the laser and Stokes radiations. In this case, Eq. 1 can be rewritten[11], neglecting the influence of χ^{NR} and Δk:

$$I_3 (x,y,L) = K^2 I_L^2 I_s \left| \int_0^L N(x,y,z) \, dz \right|^2, \qquad (3)$$

where we have accounted for possible variations of the gas number density with position and assumed that diffraction of the anti-Stokes wave is unimportant. The square root of the intensity distribution $I_3(x,y)$ at the exit boundary of the medium is thus proportional to the number density integrated along lines of constant x and y. For laser and Stokes beam intensities of the order of 1 MW/cm^2 through a 1 cm thick sample of 10% H_2 in ambient air, the anti-Stokes intensity is about 1 mW/cm^2 (at resonance, $\Delta\omega = 0$). This intensity is sufficient for exposing a high speed photographic plate in 20 ns.

For point gas concentration measurements, the pump beams are simply focused into the sample. The power collected is independent of f-number[6,11]:

$$P_a \simeq \left(\frac{2}{\lambda}\right)^2 \left(\frac{4\pi^2\omega_a}{c^2}\right)^2 P_L^2 P_s \chi^2, \qquad (4)$$

where $\lambda = 2\pi c/\omega$, (with $\omega \simeq \omega_a \simeq \omega_L \simeq \omega_s$), and where P_L and P_s are the
laser and Stokes powers respectively. P_a is generated in a small volume
about the focus, the other regions giving negligible contributions. The
focal region itself can be approximated by a cylinder with diameter
$\phi = (4\lambda/\pi d)f$ (where f is the focal length and d the common Gaussian beam
diameter in the plane of the lens) and length $\pi\phi^2/2\lambda$.

Crude calculations can be made for a typical case. Let us assume
$P_L = P_s = 1$ MW, f = 10 cm, d = 6 mm. Then with a sample containing 100
ppm of H_2 in air, we obtain 1 W of anti-Stokes power, scattered from a
volume which does not exceed 20 μm in diameter and 500 μm in length.
Spontaneous Raman scattering from the same focal volume would yield only
10^{-10} W per unit solid angle in the Stokes sideband.

III. EXPERIMENTAL RESULTS

The experimental arrangement is presented in Figure 4. It is a
standard[5,6,8,9] arrangement, with a single (transverse and longitudinal)
mode ruby laser, pumping a high pressure cell for SRS generation of the
Stokes beam* at ω_s. The cell is filled with a mixture of 80% H_2 and
20% H_e. The purpose of the H_e is to cancel the pressure shift of H_2[16].
An alternate arrangement with the ruby laser pumping a tunable dye laser
is also possible (see Ref. 4 for instance). After proper color filtering
in order to retain only the beams at ω_L and ω_s, the light can either be
sent parallel into the flow to be photographed (in which case the flow
is imaged onto the plane of the photographic plate by means of a lens)
or focused into the sample through two 8 cm focal length lenses L_1 and
L_2 corrected for chromatism between 0.7 and 1 μm; excellent spatial overlap is thus provided. Part of these beams is also focused into a
reference cell. Photomultiplier detection of the signal and reference
anti-Stokes powers, P_s and P_r respectively, is performed. As can be
inferred from a temporal analysis of both the laser and Stokes pulses
with 0.3 ns risetime equipment, the linewidths of the exciting radiations
are consistently less than 10^{-2} cm^{-1}.

In order to illustrate the detection capability of the equipment
with the focused configuration, measurements were performed on H_2
diluted in N_2, at concentrations ranging from 10 ppm to 100%. The mixture with H_2 concentrations below 10^4 ppm were produced by electrolysis

---

*In general, a tunable laser may be used to produce the Stokes beam.

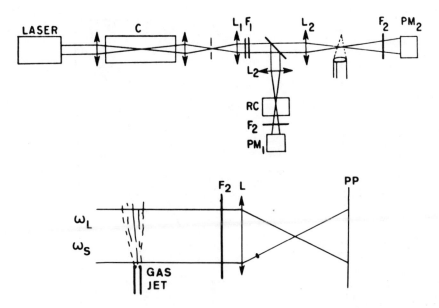

Figure 4.   Experimental arrangement. The ruby laser emits a
            diffraction-limited, single-frequency pulse. Strong
            Raman-shifted sidebands are generated in the high
            pressure stimulated Raman cell C filled with H_2.
            Only the laser and first Stokes lines must be
            retained: filters F_1 (4 mm thick Schott OG 590 +
            1 cm thick water cell) cut off the anti-Stokes
            lines and the second Stokes component after spatial
            filtering. Filters F_2 (interference filter + Schott
            VG 14 glass) admit the anti-Stokes lines generated
            in the sample (here, a flame) into photomultipliers
            PM_1 and PM_2 with S-11 photocatodes. RC is the
            reference cell. In the alternate arrangement shown
            below, the anti-Stokes radiation scattered from a
            gas cloud or jet is received on a high speed photo-
            graphic plate P.P.; lens L is not absolutely
            necessary, but improves the optical quality since
            diffraction spoils the anti-Stokes beam which would
            otherwise remain parallel.

of a 0.1 molar H_2SO_4 solution. The H_2 and O_2 thus liberated mixed with
a flow of N_2, then passed through an H_2SO_4 dessicator for removal of the
water vapor. Above 10^4 ppm, known volumes of H_2 and N_2 were mixed
directly. It is believed that the H_2 vibrational Raman lines were predominantly broadened by N_2, regardless of the O_2 or other impurity content. The resultant curve (Figure 5) is a plot on a log-log scale of the
ratio P_s/P_r versus concentration. This plot exhibits the expected slope
of 2, over most of the concentration range, but departs from this slope
at both ends.

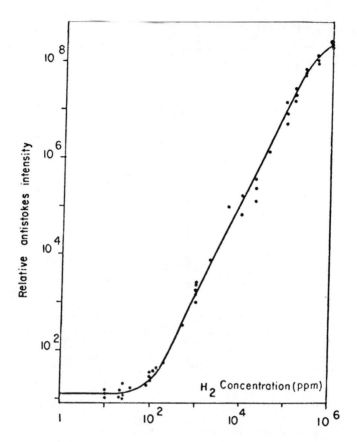

Figure 5. Normalized anti-Stokes signal versus H_2 concentration in N_2.

On the large concentration end, the combined effect of line shift and line broadening by growing concentrations of N_2 causes the observed curvature. This effect was investigated more carefully; in particular, we found that the exciting Stokes line used for the measurements of Figure 5 was upshifted by 2×10^{-2} cm^{-1} with respect to the line of pure H_2. We also verified that the addition of N_2 into the sample upshifted the line, thereby bringing it closer to resonance; this upshift reached a maximum value of 1×10^{-2} cm^{-1}, while the linewidth increased from 2×10^{-2} cm^{-1} for pure H_2 (which is in agreement with the results of Lallemand and Simova[16]) to a maximum of about 8×10^{-2} cm^{-1} for a 10:1 dilution in N_2. These results were obtained by simply scanning the line profiles by temperature tuning the SRS Stokes line. From 15°C to 150°C, an almost linear frequency sweep of -6×10^{-2} cm^{-1} could be obtained.

At the low concentration end, the background contribution from the real electronic polarizability of the $N_2 (\chi^{NR} \simeq 4 \times 10^{-18}$ esu)[6,8] gives a constant signal below 10 ppm.

This calibration curve was subsequently used to map H_2 number density profiles in a flame of natural gas premixed with air. The gas contained 75% methane and 25% ethane by volume, with traces of other gases (including H_2). The flame was horizontal, which explains the asymmetry in the H_2 distribution (Figure 6). Scans versus radial position R were made at

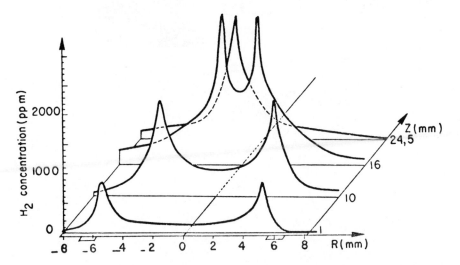

Figure 6. H_2 distribution in a horizontal natural gas flame.
The quantity R is the distance from the burner axis,
z the distance along the axis, and coordinate R the
vertical dimension pointing downwards.

various distances z from the burner; the experimental points represent averages over 5 to 10 shots. No attempts were made to correct for the influence of the temperature and of the other gaseous constituents on the line shift and broadening. The experimental uncertainty associated with the latter effect can be removed by detuning the exciting Stokes line over a number of linewidths, as mentioned above. We made such measurements off resonance on our flame, but failed to detect any appreciable change in distribution in view of the large fluctuations from shot to shot $(2^{\pm 1})$ which result from the instability of the combustion.

With the parallel configuration, and an additional 10 cm long ruby amplifier stage inserted between the laser and the SRS cell, we were able to photograph small H_2 supersonic jets in air, with a beam cross-section of 1 cm^2. Figure 7 gives an example of the picture recorded on 3000 ASA Polaroid film. The nozzle has a 0.8 mm internal diameter. The shock pattern in the stream can be distinguished.

Using the high power single mode systems presently available, beam cross-sections up to 10 cm dia. can be photographed with the same sensitivity. This is of great interest for the study of combustion, explosions, and flow mixing.

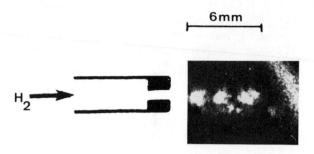

Figure 7.   Image of a supersonic H_2 jet recorded in anti-
            Stokes light.

Finally, we believe that this method may be capable of probing
rarefied gases down to 10^{-10} to 10^{-12} atm, with laser powers in the 100
MW to 1 GW range. Therefore, the method may be suitable for the remote
investigation of the upper atmosphere between 100 and 150 km altitude
from an orbiting station.

IV.   CONCLUSION

Table II presents a comparison between coherent Raman anti-Stokes
scattering and spontaneous Raman scattering; some relevant experimental
details and important properties are listed. In brief, it can be stated
that our method may not be as simple to set up as a spontaneous Raman
scattering experiment; furthermore, because of the faint, non-resonant
response of the diluent gas, the sensitivity is limited to concentrations above 10 to 100 ppm. But many orders of magnitude increase
in scattered light intensities are obtained at these concentrations,
which qualifies the method for difficult experimental situations, such
as the probing of boundary layers or the investigation of supersonic
flows between rotor blades on compressors. Furthermore, the method is
not sensitive to background fluorescence of walls or constituents in the
flow that might be excited during the measurements, because of the
directionality of the useful signal, which permits angular separation.

Table II

Comparison of spontaneous Raman scattering and coherent anti-Stokes scattering for point gas concentration measurements; p is the power scattered; N the number density; P_L, P_s the laser and Stokes powers; W_L the laser energy.

	SPONTANEOUS	STIMULATED
Experimental arrangement		
sideband	Stokes	anti-Stokes
geometry	$\Omega = 2\pi$	forward direction; same Ω as pump beams
probed volume	segment of focal zone of length x	diffraction limited focal volume
scattered power	$p \propto N \times P_L \times x$	p independent of f-number $$P \propto N^2 \times P_L^2 \times P_s$$
detectable concentrations (in partial pressure for 1 photon scattered)	10^{-6} atm with $\begin{cases} x = 1 \text{ cm} \\ \Omega = 10^{-2} \text{sr} \\ W_L = 1 \text{ J} \end{cases}$	maximum value of: -10^{-4} mole fraction x total pressure -10^{-12} atm with $P_L = P_s = 1$ GW in 10 ns

REFERENCES

1. T. Hirschfeld, E. R. Schildkraut, H. Tannenbaum and D. Tannenbaum, Appl. Phys. Letters 22, 38 (1973); R. K. Chang and D. G. Fouche, Laser Focus 8, 43 (1972).

2. G. F. Widhopf and S. Lederman, AIAA Journal 9, 309 (1971); M. Merian, Rech. Aer. 2, 85 (1972); D. L. Hartley, AIAA Journal 10, 688 (1972).

3. G. Bret, Annales Radioelectricité 22, 236 (1967).

4. I. Reinhold and M. Maier, Opt. Commun. 5, 31 (1972).

5. P. D. Maker and R. W. Terhune, Phys. Rev. 137, 801 (1965).

6. W. G. Rado, Appl. Phys. Letters 11, 123 (1967).

7. S. Barak and S. Yatsiv, Phys. Rev. A3, 382 (1971).

8. G. Hauchecorne, F. Kerhervé and G. Mayer, J. Physique 32, 47 (1971).

9. F. De Martini, G. P. Giuliani and E. Santamato, Opt. Commun. 5, 126 (1972).

10. P. R. Régnier and J.-P. E. Taran, Appl. Phys. Letters 23, 240 (1973).

11. P. R. Régnier, F. Moya, and J.-P. E. Taran, AIAA paper No. 73-702, presented at the AIAA 6th Fluid and Plasma Dynamics Conference, Palm Springs, California, July 16-18, 1973.

12. N. Bloembergen, "Nonlinear Optics", W. A. Benjamin, New York, 1965.

13. J. Lukasik and J. Ducuing, Phys. Rev. Letters 28, 1155 (1972).

14. P. P. Sorokin, J. J. Wynne, and J. R. Lankard, Appl. Phys. Letters 22, 342, 1973.

15. E. J. Allin, A. D. May, B. P. Stoicheff, J. C. Stryland, and H. L. Welsh, Appl. Optics 6, 1597 (1967).

16. P. Lallemand and P. Simova, J. Mol. Spectroscopy, 26, 262 (1968).

TARAN DISCUSSION

SMITH - What are the threshold conditions for the onset of stimulated Raman scattering in H_2?

TARAN - Typically, stimulated Raman scattering can be observed in H_2 at 20 atm and room temperature, provided the laser power exceeds 1 MW.

SMITH - Have you investigated the threshold?

TARAN - No. The work has already been done long ago. An extensive discussion has been presented by Bret (G. Bret, Annales de Radioelectricité 22, 236 (1967)). A very sharp threshold is observed in liquids because of the sudden onset of self-focusing, whereas the transition from spontaneous to stimulated scattering is quite smooth in H_2 gas; nevertheless, a number of anomalies in H_2 have been noted by Bret for single mode excitation, and by Hagenlocker, Minck and Rado (Phys. Rev. 154, 226 (1967)).

[N.B.--Editors--In response to interest in the effect of temperature on the coherent anti-Stokes scattering described by Taran at the Workshop, we asked Taran to describe application of this technique to temperature measurements. The following note gives an outline of this application.]

TEMPERATURE MEASUREMENTS BY COHERENT ANTISTOKES SCATTERING

by

J.-P. E. Taran

Equation 2 of the preceding communication, which relates the Raman susceptibility to the scattering cross section and number density, is only approximate. The number density N in the first resonant term of Eq. 2 should be replaced by the factor $N\Delta$, where Δ is the difference in probability of finding the molecule in the ground and upper states; $N\Delta$ is thus the population difference density between the two states. Equation 2 was established under the assumption that (a) the interaction does not perturb the vibrational populations and (b) all the molecules are in the ground vibrational state; these assumptions imply $\Delta = 1$.

In fact, the pump waves do perturb the vibrational population. The classical approach[1] used for the derivation of Eq. 2 gives:

$$\frac{d\Delta}{dt} \approx \frac{1}{2\hbar} \frac{\partial\alpha}{\partial q} \frac{8\pi}{c} \sqrt{I_L} \, I_S \, q.$$

Here, the vibration amplitude q is

$$q \approx \frac{1}{2m\omega_v\gamma} \frac{\partial\alpha}{\partial q} \frac{8\pi}{c} \sqrt{I_L} \, I_S \, \Delta,$$

m is the reduced mass, and α is the molecular polarizability. Taking $\Delta \approx 1$, $\gamma/2\pi c = 1\text{cm}^{-1}$, and $I_L = I_S = 1$ GW/cm^2, the vibration amplitudes are[2] of the order of 10^{-2} of those due to one quantum of excitation ($q = (\hbar/2 \, m\omega_v)^{1/2}$). They result in an initial rate of change for Δ given by $d\Delta/dt \approx 3 \times 10^7 s^{-1}$. Appreciable vibrational populations are thus produced in a few nanoseconds. This property has been used by Ducuing and coworkers for the investigation of vibrational relaxation in gases[3].

It is thus clear that care must be exercised with point density measurements, since focusing is necessary and the intensities are quite high in the focal region. Note also that an additional constraint is imposed for large laser powers P_L and large number densities ($P_L > 1$ MW, $N > 10^{19}$ cm^{-3}). Large energy transfers between the laser pump and its Raman sidebands will occur, resulting in saturation effects. Expressing the condition that the Stokes wave passing through the (diffraction limited) focus undergoes less than 10% growth due to SRS gain, one gets 2 g $P_L/\lambda < 0.1$, where the gain g is given by

$$g = \frac{(4\pi c)^2}{\hbar \omega_s^3 \gamma} \frac{d\sigma}{d\Omega} N\Delta$$

Provided the population perturbation and the saturation criteria are met, the populations in the various levels, and therefore the vibrational temperature, can be measured. To this end, one can probe simultaneously the population density difference $N\Delta_1$ between the ground and first vibrational levels and that between the first and second levels ($N\Delta_2$). This can be done with a set of appropriate laser and Stokes frequencies. Sufficient anharmonicity in the vibration is needed for a clear separation of the corresponding $0 \rightarrow 1$ and $1 \rightarrow 2$ Stokes Q-branches. Note that the method is equivalent to probing the Stokes band series near the first two maxima of its sawtooth structure, as is discussed for ordinary Raman scattering by M. Lapp in the next presentation from this workshop.

Temperature T and number density are then uniquely determined from the set of relations:

$$N\Delta_1 = N (1 - e^{-\hbar\omega_V/kT})/Q$$

$$N\Delta_2 = N (1 - e^{-\hbar\omega_V/kT}) e^{-\hbar\omega_V/kT} /Q$$

where Q is the vibrational partition function

$$Q = (1 - e^{-\hbar\omega_V/kT})^{-1} .$$

Some inaccuracy in the determination of $N\Delta_1$ and $N\Delta_2$ is caused by the complicated line profiles. One must a priori assume a value for the linewidth γ to be plugged into Eq. 2; fortunately, the various bands have similar shapes and widths[4]. Therefore, the temperature measurement which is most conveniently obtained from the ratio $N\Delta_1/N\Delta_2$ will not be strongly affected.

REFERENCES

1.   M. Maier, W. Kaiser, and J. A. Giordmaine, Phys. Rev. 177, 580 (1969).

2.   E. Garmire, F. Pandarese, and C. H. Townes, Phys. Rev. Letters 11, 160 (1963).

3.   M. -M. Audibert, C. Joffrin, and J. Ducuing, Chem. Phys. Letters 19, 26 (1973).

4.   M. Lapp, L. M. Goldman, and C. M. Penney, Science 175, 1112 (1972).

SESSION II
TEMPERATURE MEASUREMENTS; CHEMISTRY
Session Chairman: A. B. Harvey

FLAME TEMPERATURES FROM VIBRATIONAL
RAMAN SCATTERING[*]

by

Marshall Lapp

General Electric Corporate Research and Development
Schenectady, New York   12301

ABSTRACT

Raman scattering signatures are functionally dependent upon
temperature, and are therefore useful as diagnostic probes. Various
Raman scattering techniques for the measurement of temperature are
outlined here. Those methods based upon rotational molecular structure
are then briefly discussed in order to compare and contrast them with
the ones based upon vibrational structure. For flame gases, the elevated
temperatures and the multicomponent, variable composition make the vibrational scattering techniques appear to be more useful than those based
upon pure rotational scattering. Temperature measurements based upon
vibrational Raman scattering are described next, with an emphasis on
the vibrational techniques developed in this laboratory. These
techniques are based upon the spectral structure of the fundamental
Stokes vibrational band series, which consists of the ground state band
(initial $\rightarrow$ final molecular vibrational levels: $v = 0 \rightarrow v = 1$) and
the upper state or "hot" bands ($1 \rightarrow 2$, $2 \rightarrow 3$, etc.).

I. INTRODUCTION

This work is directed toward the development and implementation
of a method for temperature measurement in gases, both in thermal
equilibrium and non-thermal equilibrium conditions. The method is
based upon observation of Raman scattering vibration-rotation signatures,
and is sensitive to the vibrational and rotational degrees of freedom
of the molecular species probed.

[*]The work reported here has been supported in part by Project SQUID,
Office of Naval Research, and by the U. S. Air Force Aerospace Research
Laboratories.

The measurement of temperature is often desired, of course, for the determination of the thermal properties of the system under investigation. For example, we may wish to calculate the population of some excited state of the molecular species in question based upon Raman scattering temperature data, or perhaps we may wish to make some heat transfer estimates from temperature profiles based upon thermal equilibrium considerations. However, an estimate of the temperature is also necessary for implementation of Raman scattering density measurement schemes. This fact arises because the scattering signature is temperature-sensitive, as is the integrated scattering intensity for conditions such that excited vibrational states become appreciably populated. The temperature dependence of Raman profiles has been introduced at this meeting by Prof. Gaufres and Dr. Leonard; in this presentation we will discuss the changes in scattering profiles with temperature in a detailed fashion. The general advantages and disadvantages of Raman scattering techniques for diagnostics of combustion gases have been discussed elsewhere in this Proceedings, including the Introduction.

In the experimental work described here, we concentrate upon evaluation of the temperature from the structure of the fundamental vibrational Q-branch band series, which consists of the ground state band (i.e., scattering from the ground vibrational level to the first excited vibrational level) and the upper state or "hot" bands (i.e., scattering from excited vibrational levels to the next higher vibrational levels). (See Refs. 1-4 for further details of the vibrational Raman scattering thermal equilibrium work from our laboratory that is described here, as well as a discussion of the physical basis from which the Raman scattering profiles arise.)

II. COMPARISON OF SCATTERING TECHNIQUES

In order to illustrate the rough spectral positions and shapes of rotational and vibrational Raman scattering from a multi-component system, we show, in Figure 1, approximate rotational and vibrational spectra for N_2 and O_2 at room temperature (300°K) and at an elevated temperature (1100°K). This figure depicts the effective "spreading" of intensity with increase of temperature, and shows, for comparative purposes, the corresponding Rayleigh scattering. From this figure, we can obtain relative estimates of the strengths of the various scattering processes.   These strengths can be indicated in a quantitative fashion by differential scattering cross sections. For example, the approximate magnitudes of the differential cross sections for N_2 are, for 488.0 nm excitation:

Rayleigh	$\sim 10^{-27}$ cm^2/sr
Rotational Raman (all lines)	$\sim 10^{-29}$
Rotational Raman (strong line, including fractional population factor at room temperature)	$\sim 6 \times 10^{-31}$
Vibrational Raman (Stokes Q-branch)	$\sim 5 \times 10^{-31}$

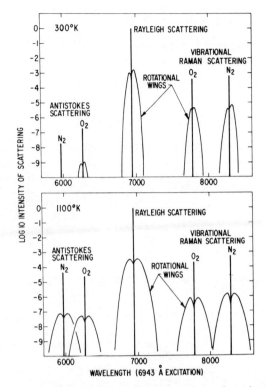

Figure 1.   Raman and Rayleigh scattering for air, drawn roughly to
            scale, at room temperature and 1100°K. To the scale of
            this drawing, ground state and upper state fundamental
            vibrational Raman bands are not distinguishable. See
            Figure 7 for details of these bands.

With this general background, we begin the comparison of Raman
scattering methods for temperature measurement by outlining several
techniques based upon rotational and vibrational scattering:

Rotational Scattering Structure:

   a.   Contour or envelope fit of band
   b.   Intensity ratio of spectral regions (including use of
            individual lines) of band (using monochromator
            or filters)
   c.   Shift of band peaks

Vibrational Scattering Structure:

   a.   Contour or envelope fit of band
   b.   Intensity ratio of spectral regions of bands (using
            monochromator or filters)
   c.   Stokes/anti-Stokes intensity ratio (using monochromator
            or filters)

    d. Shift of band peaks: Q-branch
                                   O- and S- branches
    e. Width of specific Q-branch bands, such as ground
          state band.

These methods are, for the most part, analogous to the type of temperature-measurement methods utilized in infrared spectroscopy, and will become clearer as we discuss examples of them through the use of various illustrations.

In order to discuss why vibrational Raman scattering has been used for flame diagnostics rather than rotational scattering, we consider first a general description of rotational scattering temperature measurements. (Some of this material is treated more fully in Ref. 5.) The reader will then be able to compare the rotational method to the vibrational technique, which is discussed in further detail in a later section. (N. B. Editors: Rotational Raman scattering experimental data and analyses obtained by J. A. Salzman, A. B. Harvey, J. R. Smith, and A. S. Gilbert are discussed elsewhere in this Proceedings.)

One clear attraction of rotational scattering compared with vibrational scattering is the strength of the total rotational band in comparison with the vibrational Q-branch, as is evident from a comparison of their scattering cross sections. The various methods of temperature measurement for rotational Raman scattering can be seen from inspection of Figure 2, which shows the calculated N_2 spectrum over a temperature range of 30°R (17°K) to 2000°R (1111°K). Changes in the envelope shape and shifts of the band peak intensities are clearly visible. Utilization of the intensity ratio of selected spectral regions is illustrated in Figure 3 for narrow spectral bandwidths.

The same idea is, of course, applicable to considerably wider bandwidths, as is indicated in Figure 4, where we have shown an overlay of Stokes rotational Raman scattering for N_2 and O_2 in the correct proportion for air. Here, we have schematically indicated twelve relatively broad bandpasses. Numbers 1 to 10 correspond to short wavelength cutoffs as indicated, with 527.7 nm long wavelength cutoffs. Numbers 11 to 13 correspond to bandpasses with long wavelength cutoffs less than 527.7 nm. The rotational scattered intensity S_{rot} for air passed through each of these assumed bandpasses, divided by the total Stokes Q-branch vibrational scattering S_{vib} for nitrogen, is plotted in Figure 5 as a function of temperature. (Division by S_{vib} for nitrogen is performed because of the necessity to normalize the rotational data to an easily-measured quantity which does not have a strong variation with temperature. These requirements are satisfied by S_{vib}, since this quantity is closely proportional to the vibrational partition function, which varies very slowly with temperature over the range of interest here.[6]) Thus, we see that this method of temperature measurement produces results whose sensitivity depends upon choice of the bandpass. Here, maximizing the rate of change of intensity ratio with temperature as well as the absolute magnitude of the ratio are both desirable. Depending upon the temperature range of interest, choice of filter 3 (for the lower temperatures) to filter 8 (for the higher temperatures) appears reasonable for the assumed ideal filter bandpasses.

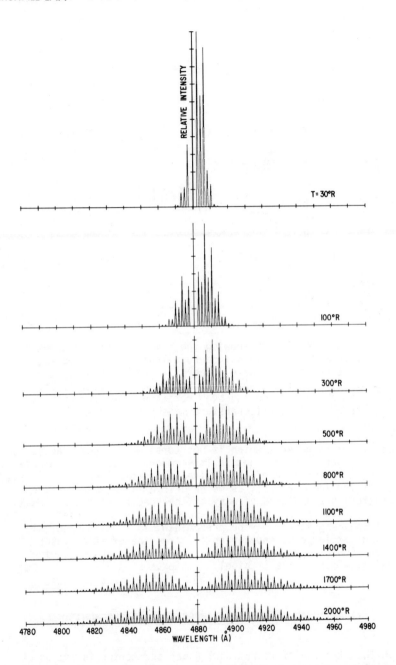

Figure 2. Calculated Stokes and anti-Stokes rotational Raman spectra
for nitrogen from 30°R (17°K) to 2000°R (1111°K). An
incident 488.0 nm laser beam was assumed, as was a triangular spectral slit function whose full width at half
maximum (FWHM) was 0.05 nm.

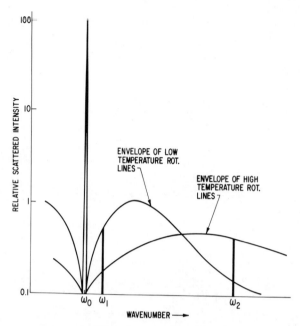

Figure 3.   Schematic diagram showing the envelopes of low tempera-
             ture and high temperature rotational Raman lines about
             wave number ω_0. Wave numbers ω_1 and ω_2 identify the
             positions of two specific rotational transitions at,
             respectively, low and high rotational quantum numbers,
             to be used in a temperature measurement scheme.

        Pure rotational Raman scattering is not considered a good candidate
for temperature diagnostics of a hot, multi-component gas whose composition is changing as a function of the desired measurement quantity (i.e.,
as a function of the temperature), as we have in a flame, because the
rotational scattering signatures of many of the species (e.g., O_2 and N_2)
overlap to a very substantial degree. In some isolated instances,
specific rotational signatures may be predominant in certain spectral
regimes (e.g., H_2 at large wavenumber shifts from the exciting laser
line, or perhaps CO_2 at small shifts), but, in general, analysis of
the rotational data for flames is highly complex.[7]

III.   VIBRATIONAL SCATTERING TEMPERATURE MEASUREMENTS

        Although the strength of vibrational Raman scattering is not as
great as that of rotational scattering, its relative freedom from spectral
interferences (See Figure 1) for different molecular species makes it more
suitable for flame diagnostics. The variation of vibrational Raman
scattering Q-branch profiles with temperature is illustrated here with
calculations made for N_2, assuming a 488 nm laser source and a monochromator
spectral slit width of 0.163 nm (corresponding, for experimental apparatus,

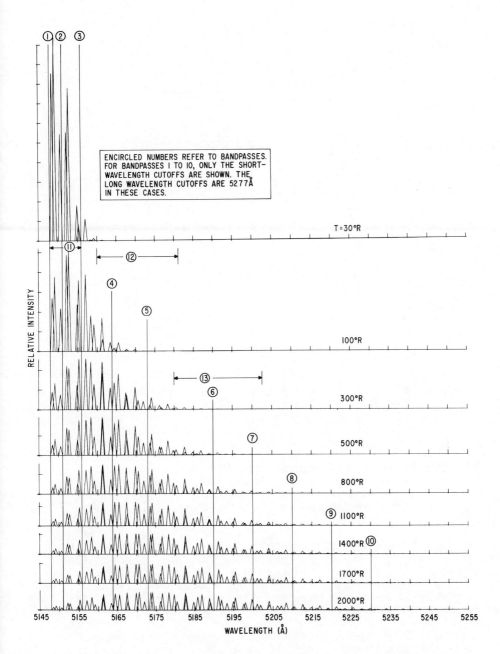

Figure 4.   Calculated Stokes rotational Raman scattering for air.
The contributions due to nitrogen and oxygen have been
plotted separately and overlaid, rather than summed.
The incident laser wavelength was assumed to be 514.5 nm.
A triangular spectral slit function was used with FWHM
of 0.05 nm.

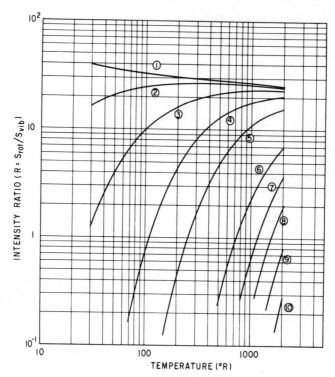

Figure 5. Ratio R of Stokes rotational Raman scattering intensity
S_{rot} for air to Stokes vibrational Raman scattering
intensity S_{vib} for nitrogen, as a function of temperature
for 10 short wavelength cutoff filter bandpasses.

to a Spex 1400-II double monochromator with 300 μm entrance and exit
slits). These calculations correspond to trace scattering, and neglect
the small depolarized contribution to the scattered intensity. Furthermore, for the spectral regime covered, O- and S- branch lines do not
make significant contributions.

In Figure 6, we see in part (A) the intensity contributions of the
individual rotational lines of the fundamental vibrational Q-branch band
series[8] (as has been discussed in these Proceedings by Prof. Gaufres), and
in part (B), the slit-convoluted spectrum that would result from an
experimental measurement. In order to review clearly the origin of the
profile shown in Figure 6, we refer to the schematic diagrams shown in
Figure 7. The top portion of this figure repeats the identification of
scattering processes shown in Figure 1 (with the exception of the
rotational sidebands on the vibrational Raman scattering) for purposes
of general orientation.

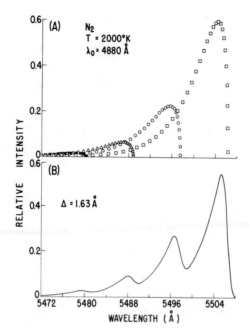

Figure 6.   Calculated Stokes vibrational Q-branch intensity at 2000°K
            for nitrogen.   (A) Alternate "strong" line intensities.
            The square data points correspond to the ground state
            band, the circular points to the first upper state band,
            the open triangular points to the second upper state
            band, etc.   (B) Triangular slit function convoluted
            profile, where Δ is the spectral slit width (FWHM).

In the center part of Figure 7, we show a detailed schematic of the
N_2 Stokes vibrational Q-branch, indicating the fact that molecular vibrational transitions from the ground state and from the various excited
vibrational levels lead to a Q-branch band series.   This arises from the
vibrational anharmonicity of the molecules, which causes the separations
of successive pairs of vibrational energy levels to be progressively
smaller.   To this level of complexity, the relative intensities of these
"stick" spectra correspond directly to the relative populations of the
molecular vibrational levels multiplied by a vibrational intensity factor
$[(v + 1)$, for the diatomic molecules considered here].

In the lower part of Figure 7, we see the added effect of (1) rotational fine structure on the vibrational stick spectrum and (2) convolution of a typical laboratory experimental spectrometer spectral slit

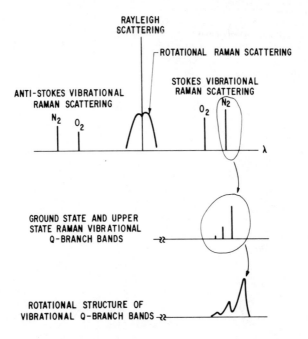

Figure 7.   Schematic of Rayleigh, rotational Raman, and vibrational
            Raman scattering. Shown in detail is the splitting of
            the vibrational Stokes Raman signal for nitrogen into a
            series of fundamental vibrational bands (ground state band,
            first upper state or "hot" band, second upper state band,
            etc.). Also shown is the spreading out of this fundamental
            vibrational band series due to rotational structure, as
            would be viewed by a typical laboratory double monochromator.

function. The fact that any of the vibrational bands shown as spikes in
the central part of Figure 7 really possess broad profiles results from
the vibration-rotation interactions of the molecules, which cause rotational lines of progressively higher angular momentum to lie at shorter
and shorter wavelengths.

The "saw-tooth" appearance of the N_2 profile shown in Figure 6B is
characteristic of many other diatomic molecules of interest for this
discussion, e.g., O_2, NO, CO, etc. A significant criterion for useful
temperature measurement schemes for these molecules can be based upon
this structure, through consideration of the separations $\delta\lambda(v+1, v)$
between successive fundamental vibrational Q-branches for no rotation
(i.e., for the J = 0 positions). Here, v is the vibrational quantum
number of the lower state for a fundamental transition, and J is the
rotational quantum number for both the lower and upper states. If the

experimental spectral resolution of the monochromator or interference filters is sufficiently good, compared to the peak separations, to be able to produce spectrally-resolved temperature-sensitive features (i.e., if the values of $\delta\lambda(v+1, v)$ are sufficiently large), then an experimentally-useful temperature measurement technique can be implemented for these molecules.

These separations are well-defined, independently of temperature, and are practically the same as the spectral intervals between peaks of the "saw-tooth" curves. Thus, for example, $\delta\lambda(1, 0)$ is the separation between the ground state $(0 \rightarrow 1)$ band and the first upper state $(1 \rightarrow 2)$ band. In Table I, values of the separations $\delta\lambda(v+1, v)$ are given

Table I

Some values for diatomic molecules of the separation of successive Stokes fundamental Q-branches, arranged according to characteristic vibrational temperature θ_v. The quantity θ_v is indicative of the population of the first excited vibrational level for any specific molecule, corresponding to a thermal equilibrium population of $(1/e)$ times the total population. Thus, it is a reasonable measure of the temperature range for which the first upper state band will approximate the ground state band in strength. (See Ref. 4.)

Molecule	$\delta\lambda(v+1, v)$ (cm^{-1})			$\theta_v(K)$
	$1 \rightarrow 0$	$2 \rightarrow 1$	$3 \rightarrow 2$	
Na_2	1.48	1.49	1.51	227
I_2	1.23	1.24	1.24	307
Cl_2	8.	8.	8.	801
O_2	23.7	23.3	23.0	2239
NO	28.0	28.0	28.0	2699
CO	26.6	26.5	26.3	3084
N_2	28.8	28.8	28.8	3354
HCl	103.6	103.3	102.9	4152
OH	165.	165.	165.	5136
H_2	233.4	231.6	229.9	5986

for a variety of diatomic molecules. We see that the commonly-found diatomic molecules (from O_2 "downward" in Table I) are suitable. We also note that the separations for light molecules (e.g., those containing H atoms) become quite large, because of their large anharmonicities. Together with their large separations of the individual rotational lines, due to their small moments of inertia which result in large rotational constants, these effects cause the overall appearance of Raman spectra for hydrogenic molecules to be "spread out" or diffuse.

Profiles of the type shown in Figure 6 for N_2 at 2000°K are shown for the temperature range of 300°K to 3500°K in Figure 8. Population

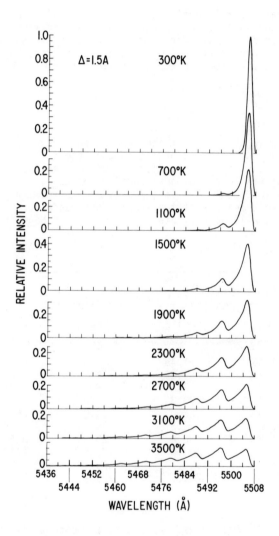

Figure 8. Calculated Stokes vibrational Q-branch intensity for nitrogen from 300°K to 3500°K.

of progressively higher vibrational and rotational states is clearly
visible here, as is the interconnection between density measurements
and temperature effects. Since the total Q-branch intensities, i.e.,
the areas under these curves, are proportional to the vibrational partition function (which rises slowly from 1.00 at 300°K to 1.05 at 1100°K,
and to 1.63 at 3500°K), we see that monitoring the gas density by
observing the total Q-branch intensity cannot be properly interpreted
(especially at the highest temperatures) without a good estimate of the
temperature. Furthermore, if monochromator entrance slit or interference
filter bandpasses are tailored to lower temperature data and then used
for higher temperature diagnostics, some of the Raman signature will be
deleted. Conversely, if the bandpasses are tailored to the higher temperature data and then used at the lower temperatures, the excessively wide
resultant bandpass could then admit extraneous signals at wavelengths
which do not correspond to the desired Raman signal, and thus degrade the
signal/noise ratio.

     The Q-branch contours for H_2 from 300°K to 3500°K are shown in
Figures 9 and 10 for the same assumed monochromator slit width as was
used for N_2 in Figures 6B and 8. They illustrate the "diffuse" nature
of the Raman profiles for light molecules, as has just been discussed in
connection with Table I. These calculated spectra are quite rich, since
each resolvable line is related to the population of H_2 molecules in a
specific vibrational and rotational state, whereas, in the case of N_2,
the assumed slit width smears the rotational intensities into a smooth
profile. (Although a fit to the N_2 profile does depend upon the rotational
population distribution function, less detailed information is contained
in such a profile than in the discrete H_2 contour.)

A. Temperature Measurements by Band Contour Fit

     From inspection of Figure 8, we see that the contour of the funda-
mental Q-branch band series varies significantly with temperature, and
therefore provides us with a method for its measurement. This method
is most easily implemented on experimentally-obtained data by normalizing
the experimental data along with theoretical curves that are calculated
at various temperatures at some common spectral position, chosen here to
be the peak of the ground state band. We then observe the degree to
which the remainder of the experimental profile (and, particularly, succeeding band peaks) fits any of the theoretical curves. The sensitivity
of this method is illustrated in Figure 11, where we show a group of N_2
Q-branches normalized at the peaks of the ground state bands.[9] They clearly
display their sensitivity to temperature in the increasing height of the
first upper state band intensity with increases in temperature - indicating
increased population of the first excited vibrational state compared with
the ground state band, and in the increasing width of the ground state band
(particularly visible here because of the normalization) and of the first
upper state band with increases in temperature - indicating increased
population of higher rotational levels as compared with lower rotational
levels.

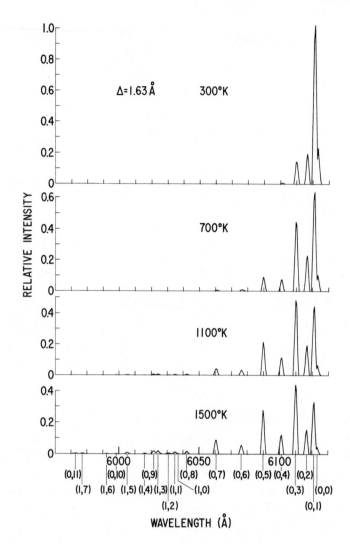

Figure 9. Calculated Stokes vibrational Q-branch intensity for
          hydrogen from 300°K to 1500°K. The vertical lines along
          the wavelength axis indicate the positions of the various
          (v,J) vibration-rotation lines, where the parenthetic
          notation corresponds to the lower-level quantum numbers.

     If we utilize Stokes Q-branch profile fits for temperature measure-
ment, we are provided with the capability of not only determining equilibrium
temperatures; non-equilibrium conditions can also be probed through an

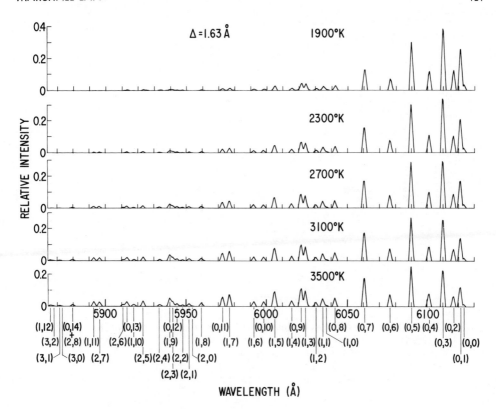

Figure 10.   Calculated Stokes vibrational Q-branch intensity for
              hydrogen from 1900°K to 3500°K.   The vertical lines
              along the wavelength axis indicate the positions of the
              various (v,J) vibration-rotation lines, where the paren-
              thetic notation corresponds to the lower-level quantum
              numbers.

analysis of an experimental profile by assigning vibrational and rotational
levels to the corresponding spectral intensities. For species such as
N_2, realistic monochromator slit widths will, in theory, smear the contours
sufficiently to cause some detailed information to be lost. However, this
is of no consequence if rotational equilibrium exists for the rotational
levels associated with _each_ specific vibrational level.

     The application of contour-fit analysis for an assumed example of
non-thermal equilibrium conditions is illustrated in Figure 12, where
we show the expected profiles for a fixed vibrational temperature T_V of

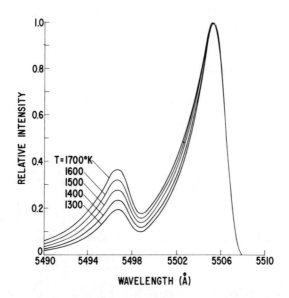

Figure 11.   Calculated Stokes vibrational Q-branch intensity for
            nitrogen from 1300°K to 1700°K, with peak ground state
            band intensity for each temperature normalized to unity.
            This set of curves illustrates the sensitivity of these
            profiles to changes in temperature.

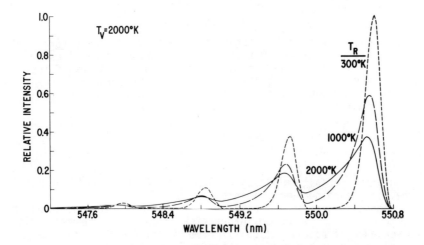

Figure 12.   Calculated Stokes vibrational Q-branch intensity for
            nitrogen with vibrational excitation temperature T_V fixed
            at 2000°K and rotational excitation temperature T_R varied
            from 2000°K to 300°K.

2000°K and for rotational temperatures T_R of 2000°K, 1000°K, and 300°K. These profiles were calculated using the formalism of Ref. 1 by assuming independent Maxwell-Boltzmann statistical distributions for the vibrational and rotational internal modes, and then by associating T_V and T_R with the appropriate terms of the expression for Q-branch spectral intensity.

A physical example related to the profiles shown in Figure 12 is the rapid rotational relaxation of a hot gas at thermal equilibrium, which can occur in a time short compared with vibrational relaxation. (Thus at 2000°K, the product of the pure N_2 pressure times the vibrational relaxation time is roughly 10^{-3} atm-sec.[10] For 10^{-3} atm (3/4 torr), this results in a vibrational relaxation time of about 1 sec, while for 1 atm, the resultant vibrational relaxation time is about 1 msec. Rotational equilibration occurs in times much shorter than these.) Figure 12 clearly shows the depletion of the molecular upper rotational levels as the temperature is reduced by depicting the strong "narrowing" of the fundamental vibrational band series.

B. Temperature Measurements by Intensity Ratios of Portions of Bands

An obvious extension (and simplification) of the temperature measurement scheme based upon vibrational Q-branch band series contour fits is the utilization of intensity ratios of selected spectral portions of the profiles. In Figure 13, we see a plot of the relative vibrational Raman scattering intensity for the filter spectral passbands indicated in the inset schematic diagram. (These filters are assumed to have spectral transmission bandpasses of rectangular shape for simplicity and clarity of calculations to illustrate the various intensities and intensity ratios. The use of narrow real filter bandpasses does not alter the conceptual results shown here.) The passband indicated by $\Delta\lambda_1$ corresponds to most of the ground state band intensity (with none of the intensity of the upper state bands), the passband indicated by $\Delta\lambda_2$ corresponds to all of the remaining Q-branch intensity, and that for $\Delta\lambda_3$ corresponds to most of the first upper state band intensity (with part of the "tail" of the ground state band). In order to obtain a sensitive indicator of temperature from these intensities, we consider the ratio R of intensity S transmitted by passband $\Delta\lambda_2$ to that for $\Delta\lambda_1$. (See Figure 14.) This intensity ratio is a rapidly varying function of temperature.

In order to refine this method to one based upon use of narrow filter bandpasses which are specifically chosen for particular molecular species (and for particular vibrational excitations), we consider the scheme outlined in Figure 15. Here, we see in the small schematic inset drawing, a slit function-convoluted Q-branch profile for N_2 at 1700°K. Drawn above this profile are indications of the bandpasses of two filters, selected to be in the spectral vicinity of the N_2 ground state and first upper state band peaks. This ratio (of the first excited state band peak intensity to that of the ground state band), shown in the main part of Figure 15 as a function of temperature, indicates the sensitivity of this method to changes in temperature for these bandpasses. In Figure 16, a more detailed plot

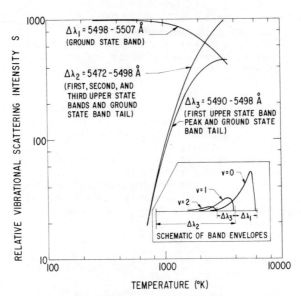

Figure 13. Calculated intensities for three spectral portions of the
Stokes vibrational Q-branch band series for nitrogen, as
a function of temperature. These spectral bandpasses are
indicated in the inset figure.

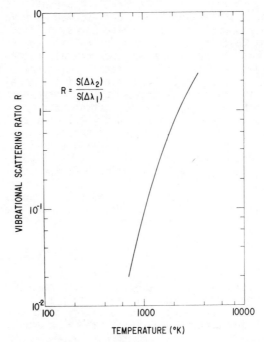

Figure 14. The ratio R of intensities for two selected bandpasses
within the Stokes vibrational Q-branch band series of
nitrogen, as defined in Figure 13.

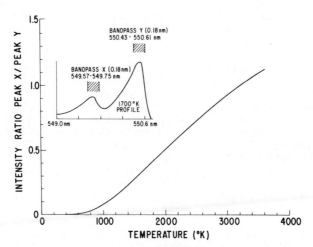

Figure 15.   Intensity ratio as a function of temperature for 0.18 nm
              rectangular bandpasses in the spectral vicinity of the
              first upper state band and ground state band for the nitro-
              gen Stokes vibrational Q-branch.

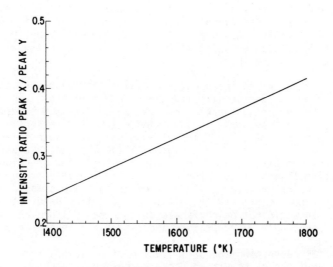

Figure 16.   Intensity ratio as a function of temperature for 0.18 nm
              rectangular bandpasses in the spectral vicinity of the
              first upper state band and ground state band for the nitro-
              gen Stokes vibrational Q-branch. The intensity ratio over
              the temperature range shown here is almost exactly linear
              with temperature.

is shown for the temperature range of 1400°K to 1800°K (similar to that
chosen for Figure 11, which depicts the temperature-sensitivity of the
overall Q-branch profile). This plot shows a useful property of this
system, namely, that the intensity ratio is almost exactly linear with
temperature over this 400°K range.

The filter spectral bandpasses can be optimized for (1) any desired
(thermal equilibrium) temperature range, (2) other experimental parameters,
or (3) the study of non-thermal equilibrium properties. Condition (1) for
optimization results from the fact that the Q-branch spectral profile
shifts with temperature, causing a filter bandpass chosen for temperature
sensitivity for a given temperature range to be displaced somewhat from
an optimally-sensitive spectral position for another range. This fact
may be restated by observing that bandpass intensity ratio curves are
nonlinear in their temperature dependence over wide temperature ranges,
and that choice of different bandpass positions and shapes (including
widths) produces intensity ratio vs. temperature curves that can have
significantly different first and second derivatives.

In Figure 17, a series of intensity ratio sensitivity curves are
shown, with one of the bandpass transmissions (for bandpass 1, which
corresponds approximately to bandpass Y of Figure 15) used as the
denominator of each of the intensity ratios. (In this figure, bandpass
X = 10 corresponds approximately to the bandpass X of Figure 15.) Calculations were performed for displacement of bandpass X to shorter wavelengths in increments of 0.1 nm, but only a selection of these curves
are plotted in Figure 17 in order to improve the clarity. Inspection of
this figure reveals the varying shapes for these sensitivity curves and the
resultant indications of usefulness for temperature measurement. For
example, curve 2 is clearly a poorer choice than curve 10 above roughly
800°K, but at temperatures somewhat below this value, curve 2 would be
a better choice than curve 10.

Condition (2) refers to those experimental areas which are not
described simply by a given sensitivity curve such as those given in
Figures 14-16. For example, suppose that we are constrained to define
a pair of filter bandpasses that would maximize temperature sensitivity
near 2000°K with some particular value of absolute wavelength tolerance
for the filter passband central wavelength. In such a case, we utilize
the fact that filters 6 to 10 all pass through roughly common values of the
intensity ratio in Figure 17, and therefore, a small wavelength uncertainty
in the filter X central wavelength for any of these filters would not inor-
dinately perturb the experimental results. On the other hand, use of
filter passband 14 would not be advisable in this case, since an uncertainty
in its central wavelength would produce a considerably larger error.

Condition (3) concerning the optimization of passbands for the study
of non-thermal equilibrium conditions, can be best understood from Figure
12, where we show the Raman signature for vibrationally-frozen, rotationally-
relaxing N_2. Suppose here that we start with a gas with vibrational
excitation temperature T_V and rotational excitation temperature T_R both
equal to 2000°K, as has already been described in Section III.A. The

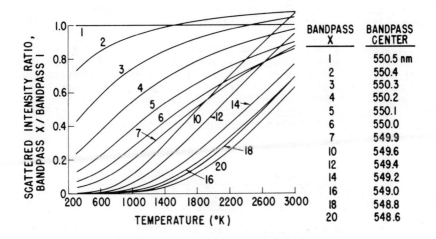

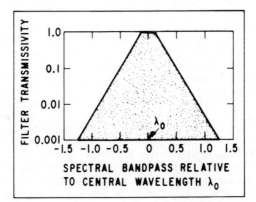

Figure 17.   Intensity ratios for selected spectral portions of the
              nitrogen Stokes vibrational Q-branch, as a function of
              temperature. The numerator for each ratio corresponds
              to the intensity observed through the various bandpasses
              whose center wavelengths are tabulated to the right of the
              figure. The denominator for each ratio corresponds to the
              intensity observed through bandpass 1. The spectral pro-
              file for each bandpass is indicated in the inset figure
              at the bottom.

Initial rotational relaxation from 2000°K to 300°K could be monitored, for example, by narrow passbands centered about 550.2 nm (detecting Raman scattering from high-lying rotational levels) and about 550.0 nm (detecting scattering from low-lying rotational levels). The ratio of these intensities coupled with the assumption of rapid equilibration among the various rotational levels associated with that particular vibrational level (the ground level) then permits us to define a rotational excitation temperature through use of Maxwell-Boltzmann statistics.

Next, suppose that we wish to monitor an ensuing vibrational relaxation (not shown in Figure 12). For that purpose, we see that narrow filters centered at 549.7 nm and 550.6 nm would monitor intensities corresponding to the molecular populations in the first excited and ground vibrational levels. The ratio of these intensities could then be used to define a vibrational excitation temperature for these levels, through use of the assumption of a Maxwell-Boltzmann distribution for these two levels.

C. Temperature Measurement by Stokes/anti-Stokes Intensity Ratios, Shifts of Band Peaks, and Widths of Bands

Probably the most widely employed technique for temperature measurement from vibrational Raman scattering makes use of the ratio of Stokes to anti-Stokes vibrational intensities. From the definitions of these types of scattering, we see that this method defines a temperature through use of the ratio of the vibrational scattering intensity from all the excited vibrational levels of the molecule to the intensity of all the vibrational levels of the molecule (i.e., including the ground level). Although this technique has been used to good advantage,[11] it suffers from two specific disadvantages. Firstly, the Stokes and anti-Stokes signatures are often in widely different parts of the spectrum, introducing difficult spectral calibration problems, and sometimes requiring different optics and detectors to optimize light collection in the two optical channels. Secondly, the Stokes/anti-Stokes method is not of use in generalized non-thermal equilibrium problems, and gives no information about rotational excitation. Besides thermal equilibrium situations, its use should be limited to non-thermal equilibrium cases for which (1) all vibrational levels are in equilibrium, but are out of equilibrium with the rotational molecular internal modes (since detailed rotational structure does not affect integrated vibrational scattering intensity data), or (2) all vibrational levels are not necessarily in equilibrium, but only two vibrational levels have appreciable population (such as the ground and first excited vibrational states). (N. B. Editors: See the presentation of G. Black in this Proceedings for consideration of case (2).) For non-thermal equilibrium situations other than these two cases, additional detailed information about the distribution of vibrational population must be available in order to interpret Stokes/anti-Stokes intensity ratio data.

Measuring the temperature from the blue shift in the peak of the vibrational Q-branch with temperature is in direct analogy to the use of similar infrared band peak shifts. In Figure 18, this shift is calculated[12] for

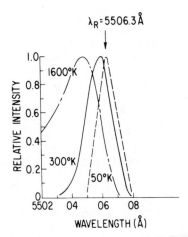

$\lambda_R = 5506.3 \text{Å}$

Figure 18.   Calculated Stokes vibrational Q-branch profiles for
nitrogen at three temperatures in the spectral vicinity
of the ground state band peaks, with the peak intensity
at each temperature normalized to unity. Here, λ_R is
the wavelength corresponding to $J = 0$ for the ground
state band. A triangular spectral slit function was
used with FWHM of 0.163 nm, as was an incident laser
source at 488.0 nm. The progressive shifts of the peaks
to shorter wavelengths is approximately proportional to
the temperature.

three temperatures in an intensity plot normalized at the peaks: (1) at
50°K, the profile is essentially the monochromator slit function; (2) at
300°K, the profile shows definite departure from the slit function shape,
and the peak has shifted 0.03 nm to the blue; (3) at 1600°K, the profile
is strongly degraded to the blue and the peak has shifted 0.15 nm. The
blue shift of the vibrational band peak is approximately given by

$$(0.348 \; \alpha_e/B_e)T - \alpha_e/4$$

to within 7% at 300°K, with the error decreasing at higher temperatures.
Here, B_e is the molecular rotational constant for rigid rotation in the
equilibrium internuclear position and α_e is a constant related to the
vibration-rotation interaction.

This Raman band peak shift is almost linear in temperature, while the
corresponding shift of the peak of a P- or R-branch ($\Delta J = -1$ or $+1$) of an
infrared band is proportional to $\sqrt{T}$. This difference arises because the
Raman peak shift is related to a $J(J+1)$ anharmonic energy level term, while
the infrared peak shift is related to a harmonic term linear in J, and
because the value of J at the peak of either type of band is proportional
to $\sqrt{T}$. However, because the infrared peak shift is related to the

rotational constant B_e while the Raman peak shift is related to the much
smaller vibration-rotation interaction constant α_e, the rate of infrared
peak shift with temperature is considerably larger. In fact, the Raman
vibrational Q-branch peak shift is not sufficiently large to provide
sensitive detection of temperature unless rather high spectral resolution
were used.

Measurement of the width of a Q-branch vibrational band profile
provides a determination of rotational excitation temperature. This method
is, of course, not suited to use of filters, and so must rely upon the
availability of a detailed vibrational scattering contour. Given that
contour, the width of a particular band is an indicator of temperature
that only uses a small part of the total data available. On the other
hand, we have already discussed the total fitting of the contour to
theoretical shapes, which makes much more efficient use of the data.
However, the width of a fundamental band can provide a convenient
estimate of temperature with far less computational effort if a vibrational profile is available.

(N. B. Editors: A. Harvey has described in this Proceedings a method
to measure vibrational excitation temperatures based upon the observation
of pure rotational Raman scattering from the ground and excited vibrational
levels. This method makes use of the interferometric "comb" method devised
by J. Barrett and described by him in this Proceedings.)

IV.   EXPERIMENTAL RESULTS

A. Experimental Equipment

The basic experimental apparatus has been described previously.[1,3]
Raman scattering was observed from hydrogen-air flames produced on a
2.5 cm diameter horizontal porous plug burner.[13] The flame was burned
horizontally into another water-cooled porous plug 2.0 cm away which was
connected to a vacuum line. A vertical laser beam from a cw argon ion
laser (~ 1 W at 488.0 nm) was focused at the center of the flame halfway
between the burner head and vacuum plug, so that the Raman scattering
test zone could be simply imaged onto the vertical slits of a double
monochromator.

The current results for N_2 correspond to use of stoichiometric H_2-
air flames.[14] Fuel-rich flames[15] were burned for the study of hot H_2 Raman
spectra. Flow rates were monitored and made steady with critical flow
orifices and regulators. The flow rates were set utilizing precision high
pressure gauges, and the flow metering system calibrated through use of
basic volume-displacement techniques.

Fine wire thermocouples[13] were made in our laboratory for independent
measurements of the flame temperature by a standard method. These thermocouples were made of 0.0005 inch-diameter wires of Pt - Pt 10% Rh,

butt-welded and coated with quartz to prevent catalytic heating. Because
of the fragility of these thermocouples, they have been replaced for most
measurements with commercially-available 0.001 inch-diameter Pt - Pt 10%
Rh units with small bead-welded ends, stretched out to the same linear
geometry as was used for the 0.0005 inch-diameter thermocouples. (Insignificant temperature differences were found by direct comparison of these
different thermocouples in our flames. Further details and schematics of
the thermocouple holders are given in Ref. 3.)

The thermocouples were moved throughout the flame with an accurate
vernier manipulator using, as a reference position locator, a finely-
machined metal cone which could be placed reproducibly on the burner
head. When the burner assembly was then placed in the test position in
front of the spectrometer, the burner could be accurately located in this
same reference position by placing it so that the laser beam just touched
the cone tip. (By observing slight attenuation of the laser beam with a
power meter at the position of the laser dump, this positioning could be
accomplished with high sensitivity.)

A schematic of the electronic signal processing apparatus is shown
in Figure 19. Data logging via paper tape for intensity and wavelength
is utilized. The wavelength data is obtained through use of an optical
incremental encoder attached directly to the wavelength drive screw of
the double monochromator.

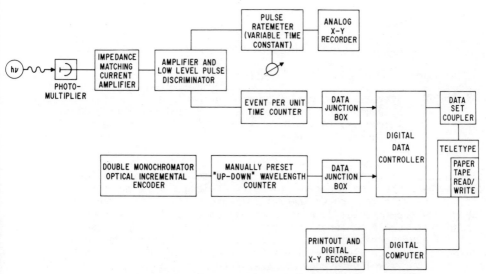

Figure 19.   Schematic of electronic detection system used with double
             monochromator. Note that the data-logging system
             simultaneously records both intensity and wavelength
             data.

B. Experimental Results for Nitrogen

The profile of nitrogen observed from the stoichiometric flame
(described in Ref. 14) at a thermocouple-measured temperature of 1575°K
(1525°K indicated temperature, plus an estimated 50°K correction for
radiative losses) is shown in Figure 20. In order to theoretically fit

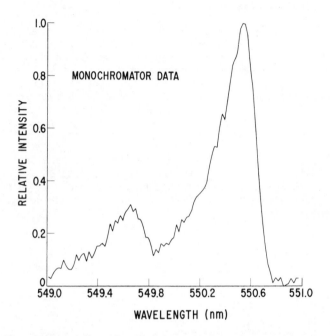

Figure 20.   Stokes vibrational Q-branch for nitrogen in a stoichiometric
              hydrogen-air flame, recorded by data logging in 0.02 nm
              intervals. A 488.0 nm cw argon ion laser source of 1 W
              intensity was used.

the experimental profile shown in Figure 20 and thereby determine the
nitrogen temperature, the experimental data wavelength axis was first
made coincident with the proper theoretical wavelength axis by manually
overlaying the experimental data on a normalized (i.e., at the ground
state band peak) set of theoretical profiles. (See Fig. 11.) These
profiles all have very similar long-wavelength edges, determined over
this temperature range almost entirely by the monochromator slit function
shape. This long-wavelength edge was therefore useful in determining
the proper absolute wavelength values here. (Current work is aimed
toward directly performing this axis adjustment, caused by nonlinearities
and non-reproducible backlash in the monochromator drive, by the computer
through use of detailed spectral calibrations made before and after each
experiment).

The next step in the determination of temperature involved a calculation based upon the ratio of intensities recorded by the monochromator
in the vicinity of the peaks of the first upper state band and the ground
state band. (See Figs. 15 and 16.) Each of these bandpasses utilized
by the computer was 0.18 nm wide, and contained ten data points. (Any
"spikes" in the data caused by experimental artifacts were first removed by
computer processing.) The theoretical ratio shown in Figure 15 (and in
more detail in Figure 16) was stored as a data file in the computer, and
the computer-determined peak ratio for the experimental data could then
be compared to this data file, resulting in a determination of temperature.

The final step in temperature measurement involved a least-square
computer curve-fitting treatment of the experimental data. The initial
assumed temperature for this procedure was that determined from the band
peak ratio method just described. The minimum least-square deviation was
then searched for by the computer as a function of temperature, and the
temperature corresponding to this minimum deviation determined thereby
to the nearest 1°K.

When we fit the experimental data to theoretical profiles, we must
smooth the data in addition to removing any "spikes." This procedure is of
particular importance, since the curve-fitting method is based upon normalization of each trial theoretical profile to the experimental data at
the peak of the ground state band. Thus, any random inaccuracy in the
experimental ground state band peak intensity can cause a substantial
distortion of the curve fitting procedure by producing a "false" normalization, with subsequent vertical stretching or squeezing of the profile.
This problem is circumvented by the averaging of adjacent data points by
the computer to produce a new "smoothed" experimental data file. As a
simple approach to this problem, we averaged odd numbers of data points
in an equally-weighted fashion. Thus, for a five-point data average at
wavelength λ, with ε equal to the spectral interval between data points,
the new intensity at λ corresponds to (1/5) times the originally-
encoded intensities at $\lambda-2\varepsilon$, $\lambda-\varepsilon$, λ, $\lambda+\varepsilon$, and $\lambda+2\varepsilon$.

In turn, this method of data smoothing has a clear shortcoming in
that it also distorts the overall profile in the vicinity of sharp, nonlinear changes of intensity. Thus, a compromise approach is dictated,
in which data smoothing is accomplished over an optimized spectral
interval. This has been done for the data shown in Figure 20, for which
three-point, five-point, and seven-point data averages were taken.
In Figure 21 is shown the five-point averaged data along with a theoretically-calculated profile computed at T = 1546°K (the temperature determined
by the peak ratio method described previously) and normalized to the peak
of the data-averaged experimental curve.

In the table contained in Figure 22 is shown the results of the
least square computer fitting procedure for the raw monochromator data,
and the three-point, five-point, and seven-point averaged data. The
temperature corresponding to the minimum deviation [i.e., T(min),
corresponding to the minimum value of Σ(deviations)] for each treatment of
the data increases here as the amount of data averaging increases. It

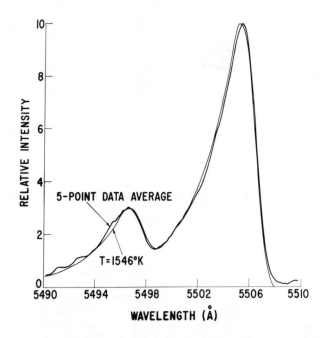

Figure 21.   Five-point data averaged Stokes vibrational Q-branch pro-
             file for nitrogen obtained from the raw data of Figure 20.
             Also shown is a theoretically-calculated profile for 1546°K
             normalized to the experimental ground state band peak
             intensity.

is easily seen that as the data averaging is increased excessively, the
spectral profile is "flattened out," resulting in an appearance closer to
that corresponding to higher temperatures.

As a working criterion for determining the optimum amount of data
averaging, the procedure chosen utilized the smallest "minimum value of
Σ(deviations)." As may be seen in the table contained in Figure 22, the
smallest value occurred for the five-point data average and, accordingly,
this was chosen as the appropriate treatment for the data. The graph
shown in Figure 22 illustrates the variation of Σ(deviations) with
temperature for the five-point data average, and indicates resultant
best fit at a value of T(min) = 1538°K.

Summarizing our present findings for the accurate determination of
flame temperature for nitrogen:

(1) Sensitivity for measurement by averaged band peak ratios is
shown in Figures 15 and 16.

(2) Sensitivity for curve fitting is illustrated by the set of
normalized curves shown in Figure 11.

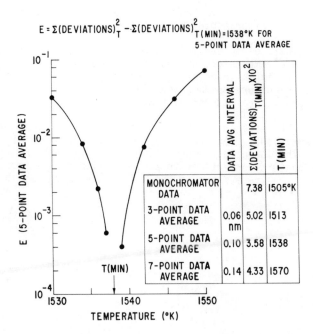

$$E = \Sigma(\text{DEVIATIONS})^2_T - \Sigma(\text{DEVIATIONS})^2_{T(\text{MIN})=1538°K} \text{ FOR}$$
5-POINT DATA AVERAGE

	DATA AVG INTERVAL	$\Sigma(\text{DEVIATIONS})_{T(\text{MIN})} \times 10^2$	$T(\text{MIN})$
MONOCHROMATOR DATA		7.38	1505°K
3-POINT DATA AVERAGE	0.06 nm	5.02	1513
5-POINT DATA AVERAGE	0.10	3.58	1538
7-POINT DATA AVERAGE	0.14	4.33	1570

E (5-POINT DATA AVERAGE)

TEMPERATURE (°K)

Figure 22.   Figure. Variation of Σ(deviations) as a function of
temperature for the five-point data averaged profile.
Table.   Summary of temperatures T(min) corresponding
to the minimum value of Σ(deviations) for the least
square profile fitting procedure, computed for the raw
monochromator data and for three cases of data averaging.
(The five-point data-averaged profile is shown in Figure
21.)

(3) For this experiment, the temperature measured by the band peak
ratio was 1546°K.

(4) The temperature measured by the best (five-point) data-averaged
computer profile fit was 1538°K, indicating agreement with the band peak
ratio method to within about 1/2% for this particular experiment.

(5) The temperature indicated by the fine wire thermocouple was
1525°K plus an approximate 50°K radiative correction, for an estimated
flame temperature of 1575°K.

This work suggests that the use of the band peak method for temperature measurements, rather than the full profile fit, is warranted for
equilibrium gases. However, the full profile-fitting method will retain
its utility for work of the highest accuracy (because of more efficient
use of the experimental data) and for investigations of non-thermal
equilibrium signatures. It will be particularly useful when neither
vibrational nor rotational equilibrium exists, for which case a deconvolution of the Raman scattering profile results in a determination of the
relative populations of the various vibrational and rotational levels.

C. Experimental Results for Hydrogen

The nitrogen data discussed in Section III. B. were taken with a
view toward accurate temperature diagnostics. The hydrogen data discussed
in this section were taken in order to investigate the different types of
Raman vibrational signature produced by a very light molecule, and were
not intended to produce results of high accuracy.

The profile for hydrogen (containing the first four rotational lines
of the Stokes vibrational Q-branch) obtained from a fuel-rich hydrogen-
air flame[15] at a thermocouple-measured temperature of about 1390°K
(including a rough 40°K correction for radiative losses) is shown in
Figure 23. Here, the trace shown as the solid curve contains the Raman

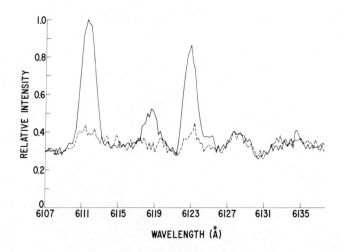

Figure 23.   Experimental Stokes vibrational Q-branch profile (solid
              curve) for hydrogen for four-times-stoichiometric hydrogen-
              air flame. The dashed curve is a subsequently-run emission
              spectrum for this slightly luminous flame. A 488.0 nm
              laser source of 1 W intensity was used, as was a triangu-
              lar spectral slit function with FWHM of 0.162 nm.

scattering data, and the dashed curve is a subsequently-run emission
spectrum of this luminous flame. The emission spectrum has been subtracted from the emission-plus-scattering spectrum in Figure 24, and
a theoretically-calculated (dashed) curve added for hydrogen at 1400°K.
Keeping the J = 3 calculated peak intensity normalized to the experimental
peak intensity, profiles were also calculated for other temperatures.
The peak values for J = 0, 1, and 2 at 1300°K, 1400°K, and 1500°K are
indicated by the appropriate horizontal lines in this figure. The
accuracy of temperature measurements for the hydrogen data shown here is
not good for two main reasons: (1) the flame is somewhat unsteady and

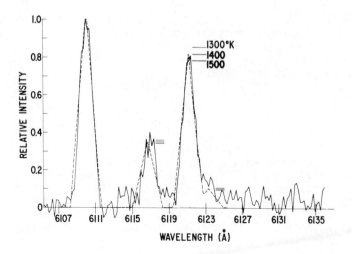

Figure 24. Experimental Stokes vibrational Q-branch profile (solid
curve) for hydrogen, obtained from the data shown in
Figure 22 by subtracting the emission spectrum (dashed
curve of Figure 22) from the scattering-plus-emission
spectrum (solid curve of Figure 22). A theoretically-
calculated profile for hydrogen at 1400°K (dashed curve)
has been added to this figure. The calculated profile
has been normalized to the experimental peak intensity
for the J = 3 line. Keeping the peak intensities
normalized for the J = 3 line, profiles were calculated
at 1300°K and 1500°K. The corresponding peak intensities
of the J = 0, 1, and 2 lines are indicated here by horizontal bars.

non-isothermal, and (2) ratios of the vibration-rotation line intensities
shown here are not particularly sensitive to the temperature over this
temperature range. Other ratios utilizing higher rotational lines are
more sensitive for this range. However, the profile presented here is
indicative of the type of data and the required treatment for temperature
estimates utilizing light molecules. The relatively wide spectral intervals between vibration-rotation lines for these molecules suggests that,
with proper choice of bandpass, interference filters could be used for
detailed excitation temperature determinations with greater ease than would
be the case for heavier molecules.

V. CONCLUSION

Methods of temperature measurement for gases based upon Raman
scattering have been reviewed. The choice of vibrational scattering

over pure rotational scattering for application to flame gases has been discussed, and data have been presented for the nitrogen and hydrogen vibrational Stokes Q-branch profiles obtained from hydrogen-air flames. Accurate determinations of temperature have been performed utilizing the nitrogen data from a band peak intensity ratio method and from a total profile-fitting procedure. The temperatures found from these methods agreed well with each other, and were also in good agreement with an independently measured temperature obtained through use of fine wire thermocouples.

We stress here that Raman scattering signatures are direct measures of the relative populations of the molecular internal modes, and, for equilibrium situations, these relative populations correspond to the fundamental definition of temperature. Thus, it is contemplated that this form of temperature diagnostics has the potential for becoming the most fundamentally accurate scheme for non-perturbing, three-dimensional measurements and, in addition, can be applied to non-thermal equilibrium systems as well.

REFERENCES

1. M. Lapp, L. M. Goldman, and C. M. Penney, Science $\underline{175}$, 1112 (1972).

2. M. Lapp, in Advances in Raman Spectroscopy, Vol. 1, ed. by J. P. Mathieu (Heyden and Son, Ltd., London, 1973) Chap. 31.

3. M. Lapp, C. M. Penney, and R. L. St. Peters, Project SQUID Technical Report GE-1-PU, Office of Naval Research (1973).

4. M. Lapp, C. M. Penney, and L. M. Goldman, Optics Comm. $\underline{9}$, 195 (1973).

5. M. Lapp, C. M. Penney, and J. A. Asher, "Application of Light-Scattering Techniques for Measurements of Density, Temperature, and Velocity in Gasdynamics," Aerospace Research Laboratories Report No. ARL 73-0045 (1973).

6. Ref. 5, pg. 109.

7. A possible exception to our conclusion that rotational Raman scattering is unsuitable for flame work is the interferometric "comb" method of Dr. Barrett discussed earlier in these Proceedings. Here, a technique based upon use of a Fabry-Perot interferometer is described that has the capability of distinguishing rotational Raman signatures for dissimilar species. Utilization of this method on a high-temperature multi-component gas mixture for diagnostics of many species with roughly similar rotational constants (i.e., roughly similar rotational line spacings) will, no doubt, be difficult, but hopefully, will be possible. (See also "note added in proof" at the end of Section III. C.)

8.   Note that only alternate "strong" lines are shown in Figure 6A,
     for purposes of clarity. Because of nuclear spin degeneracy, the
     nitrogen rotational lines alternate in intensity between "strong"
     and "weak" sets of lines, for which the statistical weights are 6
     for Q-branch lines corresponding to even values of the rotational
     quantum number J, or 3 for lines corresponding to odd values of J.

9.   The spectral position of the ground state band peaks varies with
     temperature, and is considered properly in computer-performed Q-
     branch profile fits. Here, however, when we view the profiles in
     Figure 11 over the range of 1300°K to 1700°K, the variation of
     peak position is relatively small. This is discussed further in
     Section III. C.

10. R. C. Millikan and D. R. White, J. Chem. Phys. $\underline{39}$, 3209 (1963).

11. See, for example, G. F. Widhopf and S. Lederman, AIAA J. $\underline{9}$, 309
     (1971); W. Holzer, W. F. Murphy, and H. J. Bernstein, J. Chem.
     Phys. $\underline{52}$, 399 (1970); and the presentations of S. Lederman and
     W. Kiefer in this Proceedings.

12. M. Lapp, L. M. Goldman, and C. M. Penney, General Electric Report
     No. 71-C-267 (1971).

13. W. E. Kaskan, Sixth Symposium (International) on Combustion
     (Reinhold, New York, 1957), p. 134.

14. For nitrogen data, the steady stoichiometric hydrogen-air flame
     utilized flow rates of 37.5 cc/sec H_2 and 88.8 cc/sec air, for which
     65% of the product gases was nitrogen. Since the image of the
     monochromator entrance slits at the flame scattering position was
     about 5 mm high, an estimate was made of the temperature variation
     along this zone. This was found to be about 16°K. The reproducibility
     of the thermocouple data was about ± 1/2°K.

15. For hydrogen data, a somewhat unsteady fuel-rich (four-times-
     stoichiometric) hydrogen-air flame was used (79.3 cc/sec H_2 and
     47.0 cc/sec air) for which about 51% of the product gases was
     hydrogen. The variation of temperature with position along the
     slit image was much more severe than was the case for the stoichio-
     metric flame, being roughly 110°K over a 5 mm vertical zone.
     Furthermore, the reproducibility was significantly poorer, being
     roughly ± 3°K. This flame, colored red from the emission of water
     vapor vibration-rotation bands [See A. G. Gaydon, Spectroscopy of
     Flames (Chapman and Hall, London, 1957), pp. 79, 90, and 241.]
     was subject to significantly more diffusion by the ambient atmosphere
     than the stoichiometric flame, which undoubtedly contributed to its
     less reproducible characteristics. It had, however, the virtue of a
     high hydrogen content.

LAPP DISCUSSION

BARRETT - How long did it take you to make a typical scan?

LAPP - The spectrometer was driven at 50 seconds per Å for the data shown. We've also taken traces at somewhat slower and considerably faster sweep speeds.

BARRETT - What was your signal strength?

LAPP - The number of detected photons per second for our flame source was not high -- roughly of the order of a hundred counts per second for our system, which is not optimized, I should emphasize, for sensitivity. We are using a phototube in this work, for example, with a fairly low quantum efficiency because it covers a favorable spectral regime for our overall laboratory program.

HUNTER - When you compared your experimental vibrational profile data with computer fits, the peaks of the ground state band didn't coincide in wavelength.

LAPP - Yes -- that was only an experimental artifact. The reason is that a completely accurate calibrated wavelength scale wasn't available for that data. The wavelength scale for that experiment was assumed to be linear, and the absolute wavelength magnitude determined by the best fit of the experimental data when superimposed on the theoretical plots. The currently-used system involves encoding wavelength data from an optical shaft encoder on the wavelength drive screw, which should go a long way to ensure a proper fit when combined with absolute wavelength calibration procedures.

STEPHENSON - I'm not sure that this is the right time to make this comment, but we all have signal-to-noise problems, and I would like to see more people mention either their actual counts per second, or the best they ever did, or something like that. As a starter, on a clear day with the wind from the right direction, I can get 150,000 counts per second from atmospheric nitrogen. Would anyone like to say 'I can do ten times better, or 100 times worse'?

PENNEY - What are yourconditions?

STEPHENSON - We use a 2W 4880 Å laser source, with the laser beam folded back upon itself and a mirror placed behind the scattering zone to double the scattered intensity. We use an f/0.95 collection lens and magnify the 2.5 mm-high test beam by a factor of 8 at the entrance slits of our Spex double monochromator. We also have high-reflectivity gratings and an RCA C31034A detector with a 40% quantum efficiency in the spectral region for nitrogen.

PENNEY - That is clearly a system optimized for sensitivity! If we simply scale our apparatus by the appropriate instrumental factors, we would also be in the same ballpark. But we're not there now. In fact, our "compromise" system has been set up to maximize response stability over a wide range of experimental variables, as well as resolution--and we have achieved this at the expense of sensitivity. Compared to Dave Stephenson's system, we have about two orders of magnitude lower sensitivity.

HARVEY - In regard to the noise problem, I think that we might bring up the problem of flame flickering. What good is it to enhance your signal by multipassing or some other technique, if the noise also increases proportionately? This fluctuation is caused by material of different densities moving about in the flame, producing a time-varying signal. Thus, nothing would be gained except more noise. A ratioing integrator which ratios the Rayleigh output (or better yet the total Raman signal) to the Raman output of any one spectral element would help greatly here--but, in general, why go to elaborate means to enhance your Raman signal when the dominant source of noise may be signal fluctuations?

LAPP - Yes, I agree with your basic point, but can respond in two ways. First of all, the purpose of our laboratory program is to explore the basic merits and problems of the vibrational temperature measurement scheme that I've described in my talk. For this purpose, one can produce quite steady flames, and then time-average the data as well, in order to probe the fundamentals of the method. Now--with regard to the non-steady properties of flames--the actual transient characteristics can be measured by Raman scattering if pulsed laser sources and gated detection systems are used--as has already been done by Professor Lederman. [N.B. Editors. See the presentation of S. Lederman in this Proceedings.] Obtaining data which shows the different behavior of a flame on very short or very long time scales can both be of prime importance.

GAUFRES - You have spoken about non-equilibrium conditions. There are two different possible situations: first, non-equilibrium may exist between the two degrees of freedom, rotation and vibration, but each degree of freedom is in equilibrium within itself. Second, there may be non-equilibrium within either the rotational or vibrational modes.

LAPP - Yes, I agree.

GAUFRES - But if there is no Boltzmann equilibrium for the rotational degree of freedom, the profile gives a "rotational temperature" -- if one can be defined for this case -- that would not have been the same had the equilibrium been reached. The intensity ratio of the different vibrational peaks is still a check of the equilibrium of vibrational temperature, but there is no means to check the rotational equilibrium.

LAPP - Yes -- except if sufficient resolution exists to actually observe the spectral profile in enough detail to view departures from rotational-equilibrium shapes. Particularly for light gases, such as

H_2, such resolution is easy to obtain -- but for the more commonly-encountered gases, such as N_2, it would be more difficult. For heavier gases, such as CO_2, it would seem practically impossible. But, in general, rotational equilibration occurs very rapidly in gas-kinetic systems.

HARVEY - In terms of our vibrational Raman spectrum of flowing N_2 in an electrical discharge, which you showed to illustrate non-thermal equilibrium diagnostics (see adjacent figure), our bands for each vibrational transition are much narrower than those of the flame because

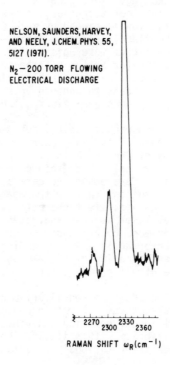

NELSON, SAUNDERS, HARVEY, AND NEELY, J.CHEM.PHYS. 55, 5127 (1971).

N_2 — 200 TORR FLOWING ELECTRICAL DISCHARGE

2270   2330
   2300   2360

RAMAN SHIFT $\omega_R (cm^{-1})$

the gas is rotationally and translationally cooler, and hence there are fewer rotational levels populated. It should be mentioned that we also looked at the pure rotational structure, which also is a measure of rotational/translational temperature, and that temperature was also very close to ambient -- so both spectra (pure rotational and vibrational) agreed fairly well.

DAIBER - How low a concentration could you measure using your technique?

LAPP - We have, we believe, quite adequate sensitivity for atmospheric flame studies in the 1500-3000°K range, for major species -- that is, for something like 1/10 normal density. This comment is based upon using our current laboratory monochromator facility, which is far from optimal for sensitivity, as we have already described. We could go down perhaps another order of magnitude by pushing our present system as well as our analysis techniques. By optimizing our monochromator facility for sensitivity, we could go down considerably further, as has been brought out by Dave Stephenson. Finally, if we go to more powerful pulsed lasers and interference filters or interferometric techniques, still lower concentrations could be measured.

DAIBER - What is the diameter and solid angle of the collecting optics that you are using?

LAPP - We currently use a photographic lens which accepts scattered light from a solid angle corresponding to f/4, and focuses the 2:1 magnified image into the Spex double monochromator at an aperture of f/8.

STORM - A question about your 3-point, 5-point, and 7-point averaging--You said that 5-point averaging gives you the best fit at this time, but is it conceivable that if you try some other number, you won't get a still better fit? It is not clear why your 5-point averaging is the best.

LAPP - We don't want to fit raw data, because even a fairly small "spike" can greatly distort the profile-fitting procedure--for example by causing the computer to pick the wrong peak position. On the other hand, we don't want to average over so many points that the real band structure is substantially distorted, because this structure carries the temperature information. Our present data-reduction technique involves least-mean-square fitting of n-point averaged data against the unaveraged theoretical curve. Thus, the final temperature value is picked by finding the minimum least-mean-square deviation as n and T are varied, with the precaution that n not be so large as to cause gross distortion. This technique leads, as n is varied, to a balance between data smoothing and distortion. The stability and consistency of our results gives us some hope that this balance is near optimum, at least for the experimental conditions we have encountered so far. However, we have not yet attempted any theoretical justification of this approach.

CHAMPAGNE - What you did to prepare your experimental profile data for curve fitting--averaging every five points--is theoretically sound and not at all arbitrary. To reply further to Eric Storm's question, it is very difficult to determine before the scan is made exactly how many data points should be included in the average. Yet this can be done objectively after the data is obtained by looking for the data-- averaging procedure--the number of points to average linearly, in this case--which minimizes the deviation in the least square analysis.

BOIARSKI - What is the highest temperature you have used in your work?

LAPP - The temperatures discussed here are limited by the fact that they correspond to stoichiometric hydrogen-air flames on a one-inch diameter water cooled porous plug burner at a flow rate of about 125 cc/ sec. It was a maximum of perhaps 1750 or 1800°K. We have produced hotter hydrogen-air flames at higher flow rates, as well as 3000°K flames with stoichiometric hydrogen-oxygen mixtures.

BOIARSKI - Do you have any data for the 3000°K flame?

LAPP - Not yet. We hope to use it to study the Raman signatures of free radicals, such as OH, which are present in significant concentrations in flames at these high temperatures.

BLACK - I wonder if in any system that you have studied, you might expect a non-Boltzmann distribution. In some flames, a lot of chemiexcitation goes on.

LAPP - We haven't seen it yet because we're working well downstream of the reactions in atmospheric pressure flames. We should see these effects if we were able to focus our laser probe beam down far enough to view a well-defined portion of the flame front. This should become quite possible as techniques are improved--particularly for low-pressure flames.

LEWIS - What work has been done on measuring the temperature for CO_2?

LAPP - We have measured the temperature of CO_2 by using the ratio of intensities of the vibrational bands corresponding to the lowest excited and the ground vibrational states. Our work was done in a cell at room temperature as well as at about 100°C. We found agreement with a thermometer to within a few percent. We simply integrated the total intensity of the Q branches to make the measurements, and as Don Leonard has also pointed out in his work on gas temperature measurements, CO_2 is a fine gas thermometer.

GAUFRES - In our work, we use vibrational Raman band analysis at a known temperature to find the ratio of spectroscopic constants α/B, which leads to determination of the vibration-rotation interaction constant α. [N.B. Editors. See the presentation of R. Gaufres in this Proceedings. In particular, refer to Section VIII, and Refs. 11 and 14 cited therein.] Unfortunately, CO_2 has poor sensitivity for this type of measurement, since α is very small due to the Fermi resonance for CO_2.

LAPP - Yes, but although it is poor for the measurement of α, the fairly sharp bands--which look almost like lines with our resolution-- make the vibrational temperature measurement techniques easier to carry out than is the case for wider bands.

GOULARD - You brought up the question of thermocouples. Can you elaborate on their relative accuracy compared to Raman probes?

CHAMPAGNE - I would like to make a comment here about relative accuracies and the need for making comparative temperature measurements with thermoelectric devices. The temperature values that Marshall Lapp obtained with a thermocouple could possibly be within 10°C of the true sampled volume temperatures. Such good accuracy is attainable with thermocouples when the system being measured is simple, and when one takes great pains to obtain accurate results.

However, in most potential engineering applications of Raman spectroscopy, the circumstances of measurement are very demanding. The

systems to be measured are seldom simple. System geometries are most
often quite complicated. Velocity and temperature distributions are
seldom even close to being one-dimensional. The emissivities of
surrounding walls and other bodies, as well as those of the fluid itself,
are usually not uniform. As a consequence, it takes much effort to
measure the mean temperature of a practical combustion effluent stream
at 1100°C to within 50°C. Inaccuracy increases with temperature, to the
extent that thermoelectric devices cannot be relied upon in many circumstances to provide accurate readings above about 1500°C. At these higher
values, temperature is usually deduced from stream composition analyses.
[See R. C. Williamson and C. M. Stanforth, "Measurement of Jet Engine
Combustion Temperature by the Use of Thermocouples and Gas Analysis,"
SAE Paper No. 690433 (1969).]

It is difficult and expensive to make accurate measurements of
temperature and other fluid stream properties with even the best available technology. This is, in fact, why laser Raman spectroscopy is
attractive as the basis for future engineering instruments. And in
order to convince engineers of the viability of this technique, good
comparative data with standard thermocouples will go a long way.

MEASUREMENT OF VIBRATIONAL AND ROTATIONAL-TRANSLATIONAL TEMPERATURES INDEPENDENTLY FROM PURE ROTATIONAL RAMAN SPECTRA

by

A. B. Harvey

Naval Research Laboratory
Washington, D. C.   20375

We show here that rotational Raman lines in vibrationally-excited gases consist of distinct components that arise from molecules in different vibrational states. These components create a line structure that can be observed and used to determine vibrational temperature.

Previously, vibrational temperatures have been determined by relative intensity measurements of Stokes-shifted Raman hot-band transitions of the type $v = 0 \to 1$, $v = 1 \to 2$, $v = 2 \to 3$, etc.,[1] by band shape analysis of such systems,[2] and by measurements of the ratio of Stokes band to anti-Stokes band intensities.[3] In this presentation we propose to extract similar information utilizing the pure rotational transitions which are one to two orders of magnitude more intense.

For a homonuclear diatomic molecule like nitrogen we may write the following expression for the rotational energy of the rotor.[4]

$$F(J) = B_v J(J + 1) - D_v J^2 (J + 1)^2 + H_v J^3 (J + 1)^3 + \ldots$$

and making use of the usual selection rule for pure rotational Raman transitions ($\Delta J = \pm 2$), we arrive at the following equation for the rotational line shifts $\Delta \nu_{rot}$:

$$\Delta \nu_{rot} = F(J + 2) - F(J) = (4B_v - 6D_v)(J + 3/2) - 8D_v(J + 3/2)^3.$$

Here, B_v is the rotational constant, and D_v, the centrifugal constant. Since D_v is usually several orders of magnitude smaller than B_v, we may assume $D_v = 0$ for this argument. Thus, $\Delta \nu_{rot}$ becomes:

147

$$\Delta\nu_{rot} = 4B_v(J + 3/2),$$

However, B_v is a function of the vibrational quantum v according to:

$$B_v = B_e - \alpha_e(v + 1/2) + \gamma_e(v + 1/2)^2 + \ldots$$

where B_e is the rotational constant at the equilibrium internuclear distance and $\alpha_e \gg \gamma_e$. If we assume $\gamma_e = 0$, then $B_v = B_e - \alpha_e(v + 1/2)$ where, for N_2, $B_e = 2.01$ cm-1 and $\alpha_e = 0.0187$ cm-1. Thus,

$$\Delta(\Delta\nu)_{v,v+1} = 4B_v(J + 3/2) - 4B_{v+1}(J + 3/2)$$

$$= 4(J + 3/2)(B_v - B_{v+1}) = 4(J + 3/2)\alpha_e.$$

Since this splitting increases linearly with J, at high rotational levels the separation of the various vibrational components is quite appreciable (see Table 1 and Figure 1); e.g., 1.6 cm-1 at J = 20 and nearly 1 cm-1 for J = 10 (just past the peak intensity) in N_2. By using a Fabry-Perot interferometer or a Fabry-Perot interferometer and a monochromator, these components can be separated and vibrational temperatures can be measured. Measurements of other pure rotational lines can yield independent information regarding the rotational/translational temperature which for some systems, particularly nitrogen, may not be identical to the vibrational temperature[1].

Table 1

Rot. Quant. # J	0	5	10	20
$\Delta(\Delta\nu)_{v,v+1}$ (cm-1)	0.11	0.49	0.86	1.6

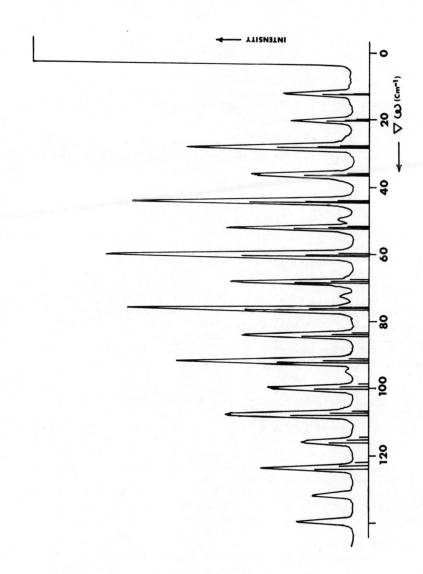

Figure 1.   Raman Spectrum of N_2 with a Schematic Display of Vibrationally Hot N_2 at High Resolution (Splitting Somewhat Exaggerated).

REFERENCES

1.   L. Y. Nelson, A. W. Saunders, Jr., A. B. Harvey, and G. O. Neely,
     J. Chem. Phys. 55, 5127 (1971).

2.   M. Lapp, L. M. Goldman and C. M. Penney, Science 175, 1112 (1972);
     see also the presentation of M. Lapp in this Proceedings.

3.   See, for example, G. F. Widhopf and S. Lederman, AIAA J. 9, 309
     (1971); W. Holzer, W. F. Murphy, and H. J. Bernstein, J. Chem.
     Phys. 52, 399 (1970); and also the presentations of W. Kiefer and
     S. Lederman in this Proceedings.

4.   G. Herzberg, Molecular Spectra and Molecular Structure I. Spectra
     of Diatomic Molecules (D. Van Nostrand Co., Inc., Princeton (1950)
     p. 553.

NOTE ADDED IN PROOF

As a result of the strongly interactive atmosphere which prevailed
at the Raman Workshop, Dr. J. J. Barrett of Allied Chemical Corp. and I
decided to carry out a joint project to test the feasibility of interferometry in the measurement of vibrational (T_v) and rotational/transla-
tional ($T_{R,T}$) temperatures in gases. Using his apparatus[5] and his technique,[6] we were successful in making measurements of T_v ($\approx$ 1840°K) and
$T_{R,T}$ ($\approx$ 375°K) in electrically discharged nitrogen using only the pure
rotational transitions (as indicated above). These measurements are in
good agreement with those measured previously[1] using the vibrational
Q-branches.

Basically, the interferometer acts like a narrow-band transmission
comb. When the spacing between the fringes is coincident with the spacing
between adjacent rotational lines, a maximum in transmission is observed.
Since the rotational constants are different for molecules in different
vibrational levels, molecules residing in higher levels will give rise
to a fringe maximum at a different comb interval (interferometer plate
spacing). T_v is then determined from the intensity ratio of the fringe
maxima for v = 0 and for v = 1. A direct fall out of these measurements
is an accurate value for $\alpha_e \approx$ 0.0178 cm^{-1} (see discussion above) which
differs somewhat from that quoted in Herzberg[4].

$T_{R,T}$ was determined from a shift in the fringe maximum (for v = 0
molecules) upon completion of the discharge circuit. The shift is caused
by an alteration in intensity distribution among the rotational transitions as the gas is heated. The shift is made apparent because of centrifugal distortion which destroys the evenly spaced nature of the
rotational lines. Another more precise method of determining $T_{R,T}$ from
these interferometric scans and the intensities of the Stokes and anti-
Stokes lines is currently under investigation.

REFERENCES

5.   J. J. Barrett and G. N. Steinberg, Appl. Optics $\underline{11}$, 2100 (1972).

6.   J. J. Barrett and S. A. Myers, J. Opt. Soc. Am. $\underline{61}$, 1246 (1971).

HARVEY DISCUSSION

BARRETT - It seems to me that your technique should work, but it certainly depends on the magnitude of the vibration-rotation constant α.

HARVEY - For nitrogen it looks attractive, because you can get a maximum intensity appearing at something like a dozen J-units (J = 12) and the separation between those lines at J = 12 is roughly 0.9 cm-1.

HENDRA - If I remember correctly, the collision broadening at one atmosphere and room temperature is on the order of 0.05 cm^{-1}. At high temperatures, isn't this broadening going to interfere with resolution of the separate peaks?

HARVEY - Well, one of the things that I didn't mention clearly is that we would like to do this work in order to go to lower pressures where pressure broadening and collisional de-excitation are minimal. This is the regime where Graham Black is working. In fact, he might tell you an amazing figure that he gets in detecting nitrogen. It's about a micron, isn't that right?

BLACK - Yes, we can detect nitrogen down to a few microns using Raman scattering. [N.B. Editors. See the presentation of G. Black in this Proceedings. The work described in this article relates to the following discussion.]

BARRETT - I didn't understand what you meant by measuring down to a few microns of nitrogen. Did you use pure rotational scattering?

BLACK - No. We looked at the Q-branch anti-Stokes scattering from nitrogen in the first vibrational level. There is no interference by ground state nitrogen. We have to measure very small amounts, because we are producing the vibrationally-excited nitrogen in a chemical reaction at low pressure.

BARRETT - What temperatures do you reach?

BLACK - We have produced nitrogen vibrational temperatures up to 1000°K by the reaction of nitrogen atoms with nitric oxide in a room temperature afterglow.

HARVEY - A few microns of nitrogen is measured by this vibrational technique, but with substantial effort. That is why we would like to explore use of rotational scattering for going to lower pressures. To return to an earlier point -- the pressure broadening that Pat Hendra referred to wouldn't be a problem even at atmospheric pressure if you to to high enough J-values. There, we're talking about line separations in nitrogen of about a wave number, which is easily resolved and is out of the pressure broadening regime at ambient conditions.

RAMAN SCATTERING AND FLUORESCENCE STUDIES OF FLAMES

by

P. J. Hendra, C. J. Vear, and R. Moss
Dept. of Chemistry, The University,
Southampton SO9 5NH, England

J. J. Macfarlane
National Gas Turbine Establishment
Pyestock, Hants, England

Presented by P. J. Hendra

ABSTRACT

Observations of resonance fluorescence and Raman scattering for hydrocarbon flames are discussed, and are related to the monitoring of composition and temperature. Results for cw and chopped argon ion laser beam excitation are presented, and proposed work with a pulsed laser source is described.

* * * * * * * * * *

The work I describe in this paper in derived to some extent from earlier high-temperature Raman work of Prof. Ian Beattie at Southampton. Beattie and his co-workers developed techniques for studying gases at temperatures in excess of 800°C using laser excitation (often with chopping techniques) and conventional commercial instrumentation (Spex Raman spectrometer). They used these methods to study equilibria in closed inorganic systems,[1] e.g.,

$$Al_2X_6 \overset{800°C}{\rightleftharpoons} AlX_3, \quad X = Cl, Br, or I .$$

Our interest resulted from a logical extension to higher temperatures and the fact that a student, Chris Vear, had worked on rotational Raman spectroscopy for his PhD. Financial support from the National Gas Turbine Establishment and the Ministry of Defense was probably stimulated by the Concorde supersonic transport program. Specifically, Concorde's early engines were smoky at low altitudes. [This was clear for all to see at the recent Biennial Air Show at Farnborough. At 15 miles range we could see the smoke, not the aircraft! Incidentally, the SAAB Draaken fighter aircraft powered by a different engine, gave no smoke--only copious brown fumes from its exhaust!] Our research was intended to fit within a long-term effort aimed at improving and controlling combustion in gas turbines in general. We were attracted to the Raman technique because of its non-disturbing and spatial resolution properties.

As a consequence of our sponsorship, we confined our attention exclusively to the hydrocarbon/oxidizing agent system. By hydrocarbon we mean methane and ethylene, although the eventual aim must be to include paraffinic volatiles such as kerosene. By oxidizing agent, we include oxygen itself but also, of course, air.

The main problem with this type of system is flame emissions, i.e., the familiar <u>Swan Bands</u>. These emissions occur particularly in the blue and green portions of the visible, are very complex, and are known to arise from the C_2 radical. The emission occurs where the stoichiometry is rich (as it is in so many real cases) and, unfortunately for us, it is particularly well endowed in the optical wavelength regions where good powerful lasers and first class optical detection systems abound; <u>viz</u>, the wavelength range 4,500 - 6,000 Å. (See Figure 1.)

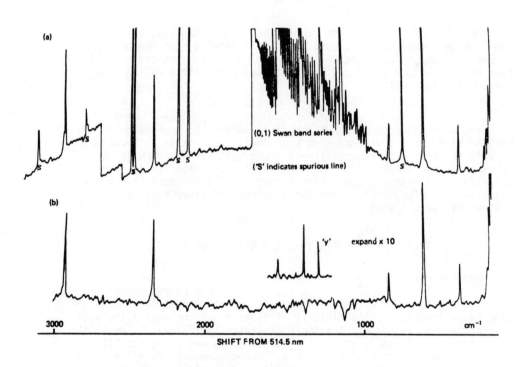

Figure 1. (a) Total emissions from domestic gas/air diffusion flame. (b) As in (a) using phase sensitive detection.

Although somewhat deterred by this emission, we attempted to record spectra and were very surprised to find strong clear spectra produced by excitation with some argon-ion laser lines (in particular 5017 or 5145 Å). (See Figure 2.) On the other hand, we saw nothing with the

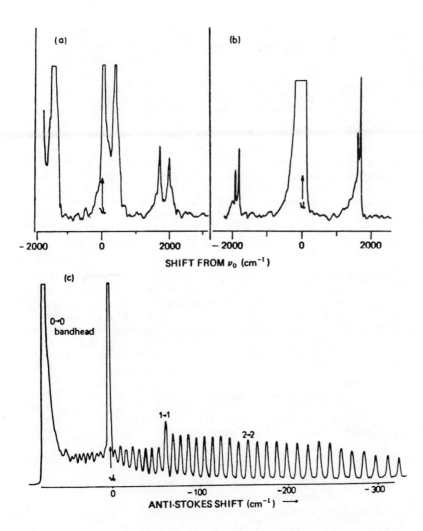

Figure 2.   (a) Low resolution resonance fluorescence spectrum of C_2 excited by λ_0 = 501.7 nm radiation. (b) Low resolution resonance fluorescence spectrum of C_2 excited by λ_0 = 514.5 nm radiation. (c) High resolution study of the origin (0,0) band of C_2 excited by λ_0 = 514.5 nm.

strong laser line at 4880 Å. Further, lean mixtures gave poor spectra.
We soon concluded that the emission is not Raman in origin but rather
fluorescence, resulting from direct excitation of the C_2 radicals
formed in the flame. (N. B. Editors. Relative characteristics that
can be used to distinguish Raman scattering and fluorescence are described
in this Proceedings in the paper by C. M. Penney and in the subsequent
discussion.)

We have been able to show that the intensity of the spectra is
high and as a consequence the sensitivity is high. We feel we can "see"
down to 10^{13} C_2 radicals/ml in the flame. If one knows the origin of
each line in the spectrum, one could use the intensities to measure the
temperature. The assignments are not clear just yet and will require
considerable effort to make them unambiguous. Making somewhat drastic
assumptions, we measure a temperature in the reaction zone of a one-
atmosphere ethylene/oxygen flame to be about 2500°K. The technique
has the same experimental advantages as the Raman process - it is non-
interfering (within reason) and provides good spatial resolution. It
is also much easier to use than the Raman measurement; thus we favor the
resonance fluorescence method as a non-sampling temperature probe in
hydrocarbon systems. We have confirmed the spatial resolution value
of the technique by recording spectra through the reaction zone of a
laboratory Bunsen flame. We confirm the rapid build-up in concentration of
C_2 species in the pre-heating zone and the minimization in the reaction
zone. Further, as expected, stoichiometry is important; as soot
information begins, the intensity of our spectra drops strongly.

Now what about the Raman scattering? We have obtained spectra
using 4880 Å excitation, very careful optical collection techniques, and
high sensitivity. In these experiments, great care was taken to collect
all the scattered light possible and as little of the flame emissions
as one could accomplish. (Resonance fluorescence is not important here,
as the 4880 Å line does not fit any transition in C_2). It was found
that, in addition to the Swan Bands we also collected some new emissions.
These are Raman bands from molecules at the laser focus.

In our next experiments, we resorted to chopped laser beams (at 700
Hz) and phase sensitive detection electronics. In this way, we recorded
only Raman bands and noise where the emission is most intense. Bands
were found as follows in many of our spectra:

H_2 — Rotational bands: 360, 592, 822 and 1040 cm^{-1}

CO_2 — 1290/1395 cm^{-1} (ν_1, ν_2 Fermi diad)

CH_4 — 2925 cm^{-1}

N_2 — 2335 cm^{-1}

O_2 — 1560 cm^{-1}

where we have given our approximate observed wavenumbers. We were
also able to use the relative intensities of bands at ν_0(4880 Å) + ν_k

and $\nu_0 - \nu_k$ to measure temperature. (N. B. Editors. See the presentations of S. Lederman, W. Kiefer, and M. Lapp in this Proceedings, and the references cited therein, for further comments on this vibrational Stokes/anti-Stokes method for measuring temperature.) In principle, one can also use band-width analysis, particularly of the N≡N stretching mode, or the relative intensities of the hydrogen rotational lines to measure temperature. (N. B. Editors. See the presentation of J. R. Smith in these Proceedings for a description of H_2 temperature measurement through use of rotational Raman scattering. For additional rotational temperature discussion, see, for example, the presentations of J. A. Salzman, A. S. Gilbert, J. W. Lewis, and M. Lapp.) Our estimates of temperature for the Raman data were only approximate but appeared to be realistic values.

The results which we have described were recorded in 1970 and 1971.[2] Attempts were then made to improve the sensitivity of the technique and to make the equipment more compact and portable so that it could be applied to large flame rigs or even gas turbine combustion chambers. To accomplish these goals, we planned a new instrument, which includes the following components:

Laser Nd-YAG solid state laser, operating quasi-continuously using Q-switching and frequency doubling. Output at 5322 Å (green) as a train of pulses at ~3.2 kHz and an equivalent continuous power of 45 mW. Pulse duration ~ 1μsec.

Monochromator Rank Precision Instruments double monochromator, 400 mm focal length gratings back-to-back. This low price instrument was drastically modified to incorporate a slit intermediate between the monochromators and a horizontal entrance slit.

Detector EMI 6256S low noise photomultiplier.

Electronics Brookdeal box-car detector.

The instrument was not successful because the laser failed to approach specification and the manufacturers went bankrupt shortly after delivery. It has been used, however, with a continuous laser, but then the box car detector (designed to discriminate temporally between random processes which are instantaneous and synchronized with the laser pulse and flame emission which is continuous) is value-less.

Our experience leads us to propose that the Nd-YAG laser may not be best for this job. We would prefer to change to the argon ion laser operating in a square wave output mode using a pulsed power supply. In this way 5-10 watts of cw power is available at a realistic price with a duty cycle of say 30 - 50%.

Further, we have recently used a most fascinating detector, the Ortec-Brookdeal 5C1 double photon counter. We favor this type of detector for flame work because it provides convenient analysis of three signals; e.g. total detector signal (A), flame background (B),

and laser power (C). The display system operates as a scaler in increments of 10^{-5}, 10^{-4}, ... 10^4, or 10^5 sec, providing measures of
$\frac{A}{C}$, $\frac{B}{C}$, $\frac{A+B}{C}$, $\frac{A-B}{C}$, or $\frac{1}{C}$. Any three data at will are then fed to the
recorder or tape punch. This capability allows some very elegant
measurements in this field.

To conclude - we have shown the feasibility of recording resonance
fluorescence and Raman spectra for hydrocarbon flames and have used
them to monitor temperature and composition. We have also tried to
indicate what we think are the next set of experiments which are
required to explore improving the sensitivities of the techniques.

REFERENCES

1.   I. R. Beattie and G. A. Ozin, Spex Speaker 14, No. 4 (Dec. 1969).

2.   C. J. Vear, P. J. Hendra and J. J. Macfarlane, J. C. S. Chem.
     Comm., 381 (1972).

     [N.B. Editors. Further observations of laser-excited fluorescence
     of radicals in flames has been reported by R. H. Barnes, C. E.
     Moeller, J. F. Kircher, and C. M. Verber, Appl. Optics 12, 2531
     (1973). Their work concerned CH fluorescence from an atmospheric-
     pressure oxyacetylene flame, and utilized a pulsed dye laser source.]

HENDRA DISCUSSION

BARRETT - Regarding the background, I'm not sure I understand why
you can't balance it out using zero suppression.

HENDRA - Because the background is far too intense. If you use
enough zero suppression to eliminate it, you have to close down the
slits and lower the gain significantly to keep the signal on scale,
or the zero suppression becoming completely unstable. When you
do this, you can't see the Raman bands at all. Added to which, the
background is not flat--that is, it contains many emission peaks. The
simplest way to do the earliest experiments was to use phase-sensitive
detection, but it isn't good enough if you want to maximize sensitivity.

LIU - What are the unique features of the data-processing system
that you use?

HENDRA - First of all, it is a pulse-counting system. I think most
of us accept that pulse counting is the way to get the most information
out of Raman spectroscopy. Another important feature is that we get
automatic background subtraction and automatic compensation for source
level fluctuations.

SMITH - What is the bandwidth of the switching unit?

HENDRA - The initial amplifier responds up to about 100 MHz, and the switching rate goes up to roughly 10^4 cycles per second which is a good deal faster than we require.

BARRETT - I am using a compact digital SSR photon counting system which I think is somewhat similar to your apparatus, in that it provides two independent counting channels, with the capability of adding, subtracting, and ratioing the outputs of the two channels.

SCHILDKRAUT - I am going to cover electronic detection apparatus in some more detail in my talk on signal processing. There are several factors to consider when information must be extracted from a noisy signal. You must differentiate between background level and noise in the background and dynamic range and signal-to-noise. All four of these things are, perhaps, problems, and just simply suppressing the DC level boosts up the dynamic range. You are talking about background signals ranging over 10^4:1 plus a real data signal of 1 percent on top of that. In my talk, I'll go into more detail, but that is basically the reason you can't use chopper electronics, and you can't use the simple photon counting electronics alone. With digital processing after the SSR photon counter, you get back some of the signal-to-noise but you can't exceed the dynamic range of the raw linear amplifier.

LAPP - I was intrigued by your comments about using fluorescence to measure CO, for example, but I'm concerned about the quenching corrections. When you are comparing the increased sensitivity of fluorescence over Raman scattering, it seems to me that one of the big advantages of Raman scattering is the absence of quenching. I wonder if it is possible to obtain a point-by-point species concentration from fluorescence in a flame, where temperature and composition vary substantially over small dimensions. It would seem that the resulting quenching corrections would be very complicated.

HENDRA - Yes, I agree, and there is another problem--absorption. The absorption through a normal Bunsen burner flame (about 1 cm in diameter) is 3 - 4%. Regarding the quenching corrections, it may be possible to obtain sufficient information by comparing fluorescence and thermocouple data. Even if there is some uncertainty in fluorescence measurements, they stand up well against other available techniques. Above all, fluorescence is a remote method of high sensitivity, and does not require material probes!

LAPP - It may provide interesting information, but if you have an unknown composition changing with space, you don't know the collision partners of the quenching. There is no way to calibrate that problem out.

HENDRA - I realize that, but inside a dirty furnace, or the combustion chamber of a gas turbine, the elegant techniques may not work. It may be that the strength advantage of fluorescence is necessary to see anything at all.

LAPP - I think many of us would argue that in a dirty system it will be even more difficult to get useful information from fluorescence. In a clean system you have a chance to calibrate out some of the problems of fluorescence.

TARAN - It seems to me that coherent Raman anti-Stokes scattering could be the right method for your problem of a dirty system. [N.B. Editors. See the presentation of J. -P. E. Taran in this Proceedings.] The scattered intensities would be many orders of magnitude higher; therefore, you would not have to worry about the emission and fluorescence of the gases and the solid particles. Take a typical example: With a gas at room temperature and 10^{-4} atm partial pressure, the stimulated Raman method would give 10 mW of anti-Stokes light for laser and Stokes beams of 1 MW power, in the focused geometry. With a 10 cm focal length f/20 lens, the active scattering volume is typically 1 mm long and 30 μm in diameter, which is very small. Spontaneous Raman scattering from the same volume yields only 10^{-10} W per steradian with a 1 MW laser beam. A gain of roughly 8 orders of magnitude is therefore readily achieved. Furthermore, the stimulated scattering is done in the same cone angle as the pump beams, which permits angular separation from the luminescence of the medium.

HENDRA - I think I could get by with 8 orders of magnitude!

MEASUREMENT OF ROTATIONAL TEMPERATURES BY RAMAN SPECTROSCOPY:
APPLICATION OF RAMAN SPECTROSCOPY TO THE ACQUISITION OF
THERMODYNAMIC VALUES IN A CHEMICAL SYSTEM

by

A. S. Gilbert and H. J. Bernstein
National Research Council of Canada, Ottawa

Presented by A. S. Gilbert

ABSTRACT

The standard expressions which describe the frequency positions and
line strengths of the rotational Raman spectra of homonuclear diatomic
gases in the rigid rotor approximation are presented. As the relative
intensities of the rotational lines are related in a fairly simple manner
to the thermal distribution of the gas molecules over the rotational
states, it is shown how it is possible to obtain a value for the temperature of a gas by measuring its rotational spectrum.

Experimental results derived from a study of the pure rotational
spectra of nitrogen and some of the practical problems encountered in
their acquisition are discussed. A scanning double monochromator with
photoelectric detection and paper chart output was used. The temperature
could be measured to about $\pm\frac{1}{2}\%$.

At total pressures of about one atmosphere and room temperature,
DCl and Dimethyl-ether vapour form a small but significant amount of a
gaseous hydrogen bonded compex $Cl\text{-}D...O(CH_3)_2$. The results of an experiment is reported in which the variation of the equilibrium constant with
temperature was followed by Raman spectroscopy. Thermodynamic values
such as the enthalpy of complexation were thereby obtained. The temperatures of the system were measured by the Raman spectroscopic method outlined above, using nitrogen gas.

INTRODUCTION

At any temperature other than absolute zero, the molecules of a gas
are distributed over a number of rotational states. For the case of a
homonuclear diatomic molecule, in the rigid rotor approximation, the

energies of the rotational states, which are each characterized by a
quantum number J, are given by

$$E = BJ(J+1),$$

where B is a constant that is proportional to the reciprocal of the moment
of inertia of the molecule about the rotational axis. For the pure
rotational Raman effect, transitions are observed between levels whose
J-numbers differ by 2 and a series of lines is obtained on either side
of the exciting line. The positions of these lines are easily derived
and are found to be given by

$$\Delta\nu = 4B(J+3/2),$$

where $\Delta\nu$ is the separation of the rotational line from the exciting line.
Stokes lines are those resulting from molecular transitions to higher
rotational levels and occur to longer wavelengths from the exciting line.
Anti-Stokes lines result from molecular transitions to lower rotational
level. Part of the Stokes pure rotational spectrum of nitrogen is
illustrated in Figure 1. The spacing between the bands is about 8 cm^{-1}
and the bands can be readily assigned to specific transitions; the numbers shown refer to the J-value of the initial state.

The general distribution of intensity over the spectrum is due to
the thermal distribution of the molecules over the rotational states.
This distribution depends on the Boltzmann factor, the degeneracies of
the states, and the nuclear spins. By taking these factors into account
it has been found[1] that the relative intensity of any line I(J) in the
pure rotational Stokes Raman spectrum of nitrogen (where J refers to the
initial level) is given by

$$I(J) = \nu^4 g(J)S(J)\exp(-BJ(J+1)hc/kT),$$

where ν = absolute frequency of the emitted line

   $g(J)$ = nuclear spin factor which is 1 for even J and ½ for odd J

   $S(J) = \dfrac{3(J+1)(J+2)}{2(2J+3)}$

      h = Planck's constant

      c = velocity of light

      k = Boltzmann's constant

      T = temperature in degrees Kelvin.

By rearrangement it is easily seen that

$$\ln \left(\frac{I(J)}{g(J)S(J)}\right) - 4\ln(\nu) = -BJ(J+1)hc/kT$$

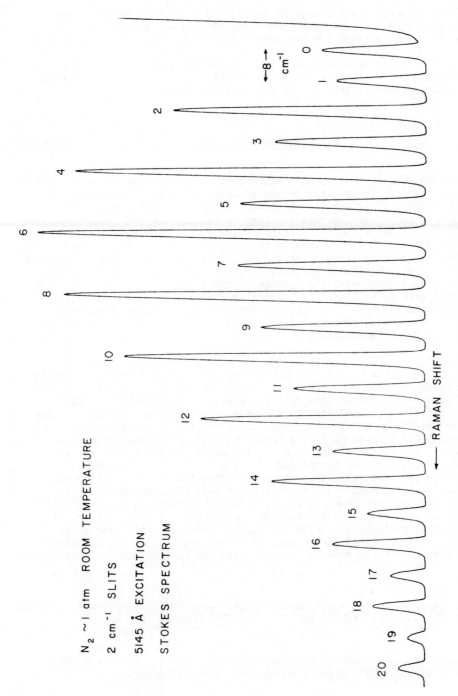

Figure 1.   Stokes Rotational Raman Spectrum of Nitrogen

If I(J) is measured for a number of bands, and the L.H.S. then plotted against J(J+1), a straight line should be obtained whose slope is equal to -Bhc/kT. Thus a value for the temperature of the gas may be calculated.

It is also possible to observe rotational transitions in association with vibrational transitions by the Raman effect, but these are found to be considerably weaker than the pure rotational lines.

EXPERIMENTAL PROCEDURES

We have recently made some measurements of temperature using nitrogen, in order to see what degree of accuracy might be obtained. A 1 metre Jarrell-Ash scanning dual monochromator with photoelectric detection to paper chart output was used to analyze the spectrum. The exciting source was a Spectra Physics cw argon gas laser yielding about 1 W of light at 5145 Å. The sample cell was contained within a multi-pass device. Spectra (from N_2 at about 1 atm.) similar to that illustrated in Figure 1 were obtained and values for the intensities (J=2-12 and 14) were provided by measuring the heights of the respective bands. The bands were considered to be completely separated from one another. A least squares fitted straight line gave 294.5°K as compared to a thermocouple and thermometer-measured temperature of 294.7°K. The standard deviation of the data points was ±0.5%.

Strictly, one should measure the relative integrated intensities (the areas under the bands) when comparing lines, as these remain constant if the shapes of the bands are deformed (as they usually are) by instrumental parameters, whereas the relative heights do not remain constant. However this is not necessary if the bands under study are all of the same shape and width, as they were assumed to be for the purposes of this experiment. It is evident that a fairly accurate determination of the rotational temperature of nitrogen may be made using the equipment described, and considering the pressure of the sample, this may be taken as equal to the translational temperature. The conditions of the experiment are, however, ideal. Difficulties arise if more than one gas is present, because each constituent will yield a characteristic set of lines, and will usually result in overlapping. An example is air. This problem can be circumvented by measuring the vibrational-rotational lines, the two sets being separated in the case of air by about 700 cm^{-1}, though of course these lines are very much weaker than the pure rotational transitions.

EXPERIMENTAL RESULTS: DETERMINATION OF THERMODYNAMIC VALUES FOR A CHEMICAL EQUILIBRIUM IN THE GAS PHASE USING RAMAN SPECTROSCOPY

When deuterium chloride and dimethyl ether (D.M.E.) are mixed together in the gas phase a small amount of a gaseous complex

$$Cl - D \ \cdots\cdots\ O \overset{\diagup CH_3}{\underset{\diagdown CH_3}{}}$$

is formed in equilibrium with the unassociated components. The specific interaction between the molecules is termed the hydrogen bond and a vast number of examples are known to occur in the physical world. Generally, the energy of interaction is an order of magnitude less than typical chemical bonds; nevertheless hydrogen bonds play a very important role in the biosphere.

The energetics of the DCl - DME interaction are of interest, and such an equilibrium may be described to a good approximation by the integration of the Van't Hoff equation

$$\ln K_p = -\Delta H/RT + \text{constant},$$

where K_p is the pressure equilibrium constant and is equal to

$$\frac{[\text{COMPLEX}]}{[\text{DCl}][\text{DME}]},$$

and

ΔH = enthalpy of complexation

R = gas constant

T = temperature in degrees Kelvin.

By observing the change in K_p with temperature it is possible to calculate ΔH and then by use of standard thermodynamic relations to obtain values for the entropy and free energy changes.

We chose to follow the changes by monitoring the pure vibrational Raman band of the unassociated DCl which scatters at about 2090 cm^{-1} from the exciting line. (There are actually two closely overlapping DCl bands due to the presence of two isotopes of chlorine.) The stretching vibration of the complexed DCl scatters as a broad and intrinsically much weaker band centered at around 1885 cm^{-1} as illustrated in Figure 2. (Gas pressures and instrumental parameters used to record the spectrum shown on Figure 2 differ from those used on the main experiment.) Because of the experimental configuration, nitrogen gas was mixed with the constituents in order to act as an internal standard, and the integrated intensity of the pure vibration band of the unassociated DCl (the Q branch) was thus compared with the integrated intensity of the pure vibration band of the nitrogen, which scatters at about 2330 cm^{-1} from the exciting line, over a range of temperatures. It was our original intention to measure the temperatures of the system by means of a thermocouple but instead we decided to make the nitrogen serve a dual purpose by using its pure rotational spectrum to measure T. Some interference was received from the pure rotation lines of the DCl and DME but this was not serious as the former was only present in a relatively small quantity (50 torr DCl, 750 torr DME, 300 torr N_2) while the latter's rotational transitions

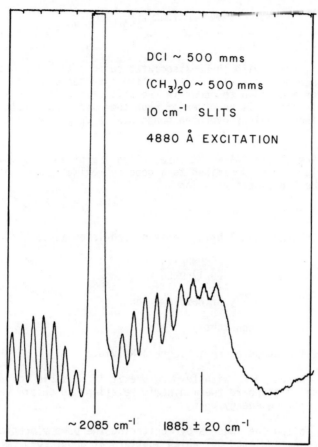

DCl ~ 500 mms

$(CH_3)_2O$ ~ 500 mms

10 cm^{-1} SLITS

4880 Å EXCITATION

~ 2085 cm^{-1} 1885 ± 20 cm^{-1}

Figure 2.   Vibrational Raman scattering from unassociated DCl
(~ 2085 cm^{-1}) and from complexed DCl (~ 1885 cm^{-1}).

are very closely spaced and were only visible as a change in background
intensity.

The results of the experiment are illustrated in Figure 3, ΔH being
calculated from the slope of the graph. The whole experiment is a
demonstration of what Raman spectroscopy can achieve. Of course, it is
possible to use other techniques to monitor the system, such as infrared
spectrophotometry. However the infrared absorption spectra of gases are
generally dominated by rotational features which are often quite extensive. Unassociated hydrogen halides do not yield a pure vibration band
(Q-branch) at all and thus in mixtures recourse may only be made to the
rotation-vibration wings (P-and R-branches). A further complication
arises in hydrogen bonded systems in that the vibration band of the complexed species (such as the D-Cl) absorbs as a broad and intrinsically
very intense band that usually overlaps with the unassociated molecular
band. By contrast, the dominant features of Raman spectra are the
central Q-branches.

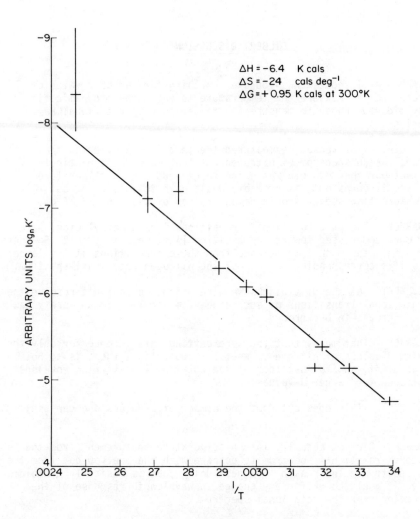

Figure 3.   Plot of temperature variation of equilibrium constant, used to calculate the enthalpy of complexation (ΔH) from the slope.

REFERENCE

1.   G. Herzberg, Molecular Spectra and Molecular Structure I.   Spectra
     of Diatomic Molecules, 2nd ed. (D. Van Nostrand Co., Inc.,
     Princeton, 1950) p. 128.

GILBERT DISCUSSION

LAPP - As I understand, you are searching for the best value of ΔH,
and therefore you measure the temperature to get a thermodynamic fit.
But why did you choose to measure the temperature by rotational Raman
scattering?

GILBERT - The reason we measured the temperature by using the
rotational Raman spectrum of nitrogen was that we weren't particularly
happy that our thermocouple was going to give us a good temperature
value. As it turns out, it probably would have in this case, but this
is an interesting spectroscopic demonstration of what one might do.

HARVEY - The pure rotational transitions from dimethyl ether were
pretty much compressed toward the exciting line, were they not? So there
really wasn't too much interference from these transitions at the larger
Raman shifts corresponding to most of the nitrogen rotational spectrum.

GILBERT - At the slit widths we were using, we couldn't resolve the
pure rotational transitions of such a heavy molecule. So we merely saw
a large increase in background.

HARVEY - That means that you were getting interference for only the
first few J-values of nitrogen. What I'm suggesting, then, is to not
worry about it, and to just look at the rotational scattering envelope
for nitrogen for larger J-values.

GILBERT - This does cut down the number of J-values one can use,
however.

BLACK - Did you actually make a temperature measurement from the
anti-Stokes rotational spectrum to compare with the one you got from the
Stokes spectrum? The reason I suggest doing that is basically twofold:
firstly, it would check for any change in wavelength response of the
spectrometer over the rotational spectrum.

GILBERT - We have assumed this to be constant over the range. We
have calibrated our spectrometer with a standard tungsten lamp and within
the other errors of the experiment, the response of the spectrometer is
effectively constant.

BLACK - The second thing would be to see whether the ν^4 term can be taken as a constant and if so whether you will get a straight line.

GILBERT - Originially we didn't use the ν^4 term. But with regard to this problem, you can't really distinguish whether you get a straight line or not. Over the range we were observing there is about a 5% change in ν^4, but when you take logs, it goes down to about 1%. So it changes the value for the absolute temperature by about 1% if you ignore the variation of the ν^4 factor.

TRANSIENT FLOW FIELD TEMPERATURE PROFILE
MEASUREMENT USING ROTATIONAL RAMAN SPECTROSCOPY

by

J. R. Smith*

Sandia Laboratories, Livermore, CA

ABSTRACT

The application of a pulsed laser and low-light-level television system for obtaining the frequency and spatial distribution of pure rotational Raman scattering from a transient gas flow is presented. The system has a temporal resolution of less than one millisecond and spatial resolution better than 0.1 millimeters. Examples of temperature determinations of hydrogen and deuterium are given. Plans are to use the system to take temperature profiles across a transient gas jet.

I. INTRODUCTION

There is a significant need for fast, non-perturbing temperature measurement techniques in transient cold (non-radiating) gas flows. Pure rotational Raman scattering has been selected as a profitable approach because Raman scattering does not perturb gas flows if power densities are low enough to preclude stimulated Raman scattering and ionization breakdown. Barrett and Adams[1] have shown the requirements for scattering in ultra-small gas volumes. Bridoux and Delhaye[2] have demonstrated Raman spectrum collection by means of low light level television. Widhopf and Lederman[3] have had considerable success in making fast scattering measurements using a pulsed ruby laser. Salzman, Masica, and Coney[4] have examined many of the problems of rotational Raman temperature measurements. Drawing from these experiences and adding the requirement of spatial resolution led to the development of the system described herein.

---

*Work done in partial fulfillment of requirements for Ph.D, University of California at Davis, Mechanical Engineering Department. This work was supported by the United States Atomic Energy Commission, Contract Number AT-(29-1)-789.

II. APPARATUS AND TECHNIQUE

Using a small cylindrical cell to contain various diatomic gases, a long pulse (500 to 1000 microsecond duration) of a few joules from a Korad K-5 ruby laser (694.3 nanometers) is passed through the gas. The laser beam is apertured to a few millimeters diameter and is focused into the cell to form a cylindrical scattering column 0.05 millimeter in diameter with a length of 15 millimeters. The Raman-scattered light is collected perpendicular to the laser beam by crossed cylindrical lenses of 50 millimeters and 100 millimeters focal length. Such an arrangement allows imaging of the entire scattering gas column within the f/6.3 acceptance cone of the 3/4-meter Jarrell-Ash single pass spectrometer. The gratings used are 300 groove/mm or 1800 groove/mm Bausch and Lomb gratings, especially selected for low ghost character-istics. The low density grating is used for hydrogen and deuterium while the high density grating is used for oxygen and nitrogen.

No exit slit is used and the entire rotational spectrum is focused onto an RCA 8606 image intensifier tube. This is a three-stage electro-statically focused intensifier with a brightness gain of 105,000. Only the central 25 millimeters of the 40 millimeter diameter intensifier is used to avoid excessive pin cushion distortion. The output of the image intensifier is fiber-optic coupled to a Westinghouse EBS (silicon tube similar to the SEC) 606 low light level television camera. Figure 1 shows the general arrangement of the system.

The data is recorded on a Sony 3650 video tape recorder. The video tape is then digitized for data reduction. One television scan line (as displayed on an oscilloscope) of the rotational Raman spectrum of oxygen is shown in Figure 2. The digitized spectrum of each line of the television scan is fit with a least squares routine that includes system response as a function of wavelength and data weighting proportional to the amplitude of each rotation level. At present only the peak amplitude is used instead of the more appropriate integral of each rotation level. Examples of static gas temperature measurement fits for hydrogen and deuterium at 294°K are shown in Figures 3 and 4, respectively. The hydrogen fit of 285°K is 2.9 percent low. The deuterium fit of 319°K is 8.5 percent high. It is hoped that more pre-cise spectrometer calibration and improvements in data reduction techniques will reduce the temperature error.

The major difficulty in such a system is stray light from the very intense Rayleigh line (about 1000 times more intense than the rotational Raman signals). The Rayleigh line itself is attenuated by a strip of neutral density (ND3) filter placed over the center of the face of the image intensifier tube.

The use of the television camera allows spatial resolution of temperature variations along the path of the laser beam. Since the 15 millimeter high spectrum is scanned by 315 of the available 525 camera scan lines (standard commercial scan), the spatial resolution is about 0.05 millimeters. The spatial resolution is adjustable by cylindrical

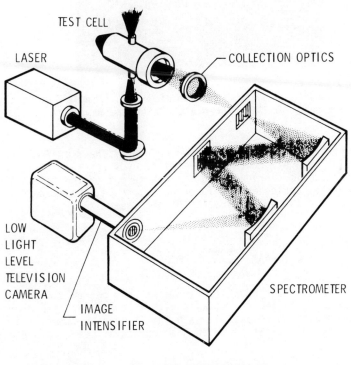

TEST CELL

LASER

COLLECTION OPTICS

LOW
LIGHT
LEVEL
TELEVISION
CAMERA

IMAGE
INTENSIFIER

SPECTROMETER

SYSTEM RESPONSE TIME __ 500 MICROSECONDS
REPETITION RATE _____ 30 PULSES PER SECOND
SPATIAL RESOLUTION _____ ADJUSTABLE FROM 0.025 TO 1mm

Figure 1.   Rotational Raman temperature profile measurement optics.

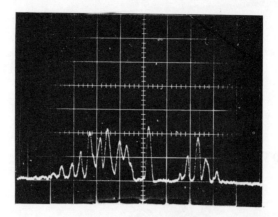

Figure 2. Rotational Raman spectrum of oxygen (single line of TV
scan). An ND3 filter attenuates the Rayleigh line.

lenses at the entrance or exit of the spectrometer. The scanning
process is continuous and requires 33.3 milliseconds for each frame.
However, the spectrum is stored on the silicon target of the television
tube in the same time period as the laser pulse; hence, the temporal
resolution depends on the pulse duration of the laser.

III. FUTURE PLANS

The present laser has a repetition rate of only two pulses per
minute. It is planned to acquire a laser with a repetition rate of
30 pulses per second to be compatible with the television camera framing
rate. Also the data digitization is in the process of being completely
automated.

An analysis of the temperature distributions within the cylindrical
gas cell when quickly filled by H_2, D_2, O_2, or N_2 in a jet from a
circular orifice will be completed shortly. This technique for
obtaining temperature profiles is ideally suited for use in boundary
layer flows.

REFERENCES

1.   J. J. Barrett and N. I. Adams, III, J. Opt. Soc. Am. 58, 311 (1968).

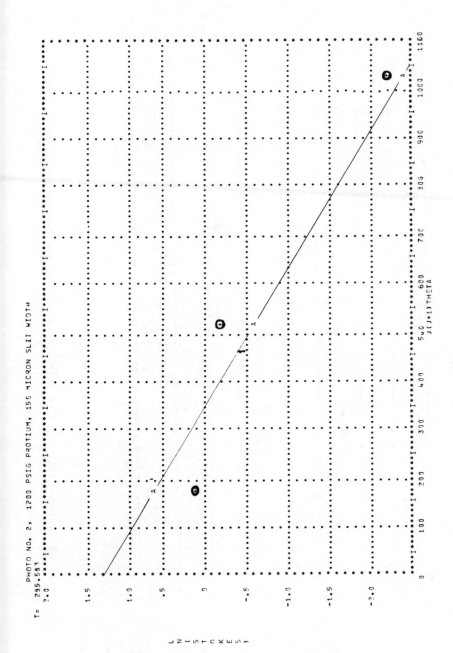

Figure 3. Hydrogen rotational line intensity plot used for temperature determination.

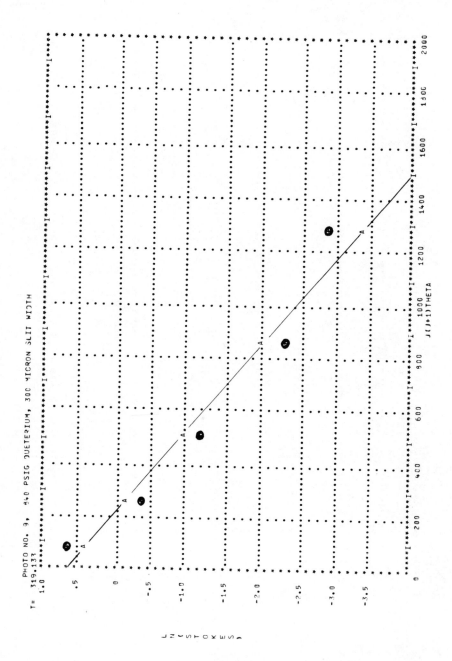

Figure 4.   Deuterium rotational line intensity plot used for temperature determination.

2.   M. Bridoux and M. Delhaye, Nouv. Rev. d'Optique 1, 23 (1970).

3.   G. F. Widhopf and S. Lederman, "Specie Concentrations Measurements
     Utilizing Raman Scattering of a Laser Beam," Polytechnic Institute
     of Brooklyn, Department of Aerospace Engineering and Applied
     Mechanics, PIBAL Report No. 69-46 (1969).

4.   J. A. Salzman, W. J. Masica, and T. A. Coney, NASA TN D-6336 (1971).

SMITH DISCUSSION

SCHILDKRAUT - I would steer away from a better image intensifier;
what it will do is decrease your already small dynamic range and compress
it to the point where you may get 2-to-1 instead of the 6-to-1 that you
probably now have. A gain of a million in the intensifier up ahead of
your vidicon photocathode, which also has compressed dynamic range,
will give you enough dark counts and scintillation due to the intensifier
photocathode to possibly obliterate your signals. Remember, the
intensifier has a phosphor with very limited dynamic range (10:1) and
you are not photon counting; you are using a standard TV raster which
means integrating each TV resolution element for 0.12 microseconds.
You really can't go in the direction of more early stage gain, if you
have a choice - go with a bigger laser.

SMITH - Yes, that is really the choice. A new laser does promise
20 pulses per second, non-Q switched, with 5 to 10 joules per pulse,
which means that with this system, real time transient temperature
profiles for boundary layers can be obtained. If we can get the
resolution up a little bit, we can also make measurements in low
density wind tunnels.

KIEFER - I did not understand why you are using collection optics
which consist of two crossed cylindrical lenses. Did you get any
improvement compared to high speed circular optics?

SMITH - In the vertical direction, I wanted a magnification less
than one, in order to define the scattering volume in which we are
interested. But in the slit-width direction, I wanted to maximize
the collection efficiency, and that means going to high magnification
in this direction.

BARRETT - Have you considered the use of a channel plate multiplier
instead of a three-stage electrostatic image intensifier?

SMITH - I have one on order and it is 3 months overdue. I see a
significant improvement in terms of resolution right now. When you
couple things together, it is not clear exactly how the resolution is
going to add, but there are ways of estimating that. The image
intensifier is supposed to resolve 28 line pairs per millimeter; I

measured 18. The camera supposedly has 18 to 23; I measured it at more
like 14 line pairs per millimeter. The Bendix channel plate units
supposedly resolve 40 line pairs per millimeter, but if you want to
buy the high gain unit, it is relatively expensive compared to the
$3,800 RCA tube; $20,000 for the one with the gain of 50,000. That
is a big price difference, for a small difference in resolution.

BARRETT - Would it improve your dynamic range though?

SMITH - Well, I will have to discuss this dynamic range problem with
Dr. Schildkraut. But my indications right now are I am limited in
dynamic range by the television camera and not by the image intensifier.
You can control the gain of the intensifier relatively easily by changing
the voltage, since it is a low voltage system and of reasonably low
capacity.

BARRETT - Is this a silicon television tube?

SMITH - Yes, a silicon intensified vidicon.

LIU - Is there any reason to use the long pulse non-Q switched
lasers? I wonder if it has anything to do with the limitation of the
dynamic range or the scanning rate of your instruments.

SMITH - No, it is not limited by either. It is limited by the
fact that if you increase the power density by compressing the pulse
time, there is a possibility of getting stimulated scattering or gas
breakdown, which is a disaster for the detector.

BOIARSKI - What was your slit width?

SMITH - For hydrogen, you can have a slit width of several hundred
microns because its spectrum is so widely dispersed. The oxygen data
was obtained with slits of 30 to 40 microns. A great deal of discussion
has taken place about line broadening phenomena at high pressures.
I do have a single line scan of oxygen taken at an absolute pressure of
1500 psi with all the oxygen lines very well resolved, almost down to
the base line. This says that the pressure broadening effect is not
nearly as bad as some people think it is.

TARAN - It depends on the gas.

SMITH - Yes, well it depends on the B-value, the gas, and on many
other things.

TARAN - Nitrogen has an interesting behavior; its line width decreases when the pressure is raised (E. J. Allin, A. D. May, B. P.
Stoicheff, J. C. Stryland, and H. L. Welsh, Applied Optics $\underline{6}$, 1597
(1967)).

LOW TEMPERATURE MEASUREMENTS BY
ROTATIONAL RAMAN SCATTERING

by

Jack A. Salzman

National Aeronautics and Space Administration
Lewis Research Center, Cleveland, Ohio

ABSTRACT

In this presentation, concepts are developed for the use of rotational
Raman scattering for gas-phase temperature measurements. Comparisons between experimental and theoretical air spectra are given, as are analyses
and experimental data related to the measurement of temperature by utilization of ratios of rotational line intensities.

*   *   *   *   *   *   *   *   *   *

Among the many suggested applications of the Raman scattering of
laser light is the measurement of gas temperatures. The intensity distribution of the Raman spectrum of a gas in thermodynamic equilibrium is
a function of the local gas temperature and thus, temperature data can
be extracted from an analysis of that spectrum. The idea of this application of Raman scattering was propounded early by Cooney,[1] was later
expanded by Pressman[2] and has since been developed by a number of
investigators.[3,4,5,6] The methods developed for temperature measurements have employed a number of ways of analyzing the Raman spectrum
including 1) a complete line-by-line intensity analysis, 2) a comparison
of the envelope of all the lines, and 3) an examination of the ratio of
the intensities of two different spectral regions.

The rotational Raman spectrum has particular application to measurements of temperatures up to several hundred degrees Kelvin because in this
temperature regime it has the strongest scattering intensity and its
relative line intensities vary most acutely with temperature. Analyses
of the relative line intensities of an ideal line spectrum of a linear,
diatomic molecule such as oxygen or nitrogen have been performed many

times (e.g., Refs. 3, 4, and 5). Because these spectra result from molecu-
lar-energy-level transitions from the initial state j to j + 2 and j - 2,
the line intensity I_j is determined by two component equations representing
the Stokes and anti-Stokes scattering respectively. For example, the
line intensity of the Stokes half of the spectrum (j → j + 2) is

$$I_j^S = \frac{C(j+1)(j+2)}{2j + 3} \frac{\Theta}{T} (\nu_0 + \Delta\nu_j^S)^4 \exp[- j(j+1) \frac{\Theta}{T}]$$

where $\Delta\nu_j^S = \frac{4k\Theta}{hc} (j+3/2)$.

The constant C is a proportionality constant which includes a nuclear
spin degeneracy factor, the rotational characteristic temperature of
the scattering gas is given by Θ, and T is the gas temperature.

    The line spectrum represented by such calculations is illustrated
in Figure 1 for gaseous nitrogen at three temperatures. At the lower

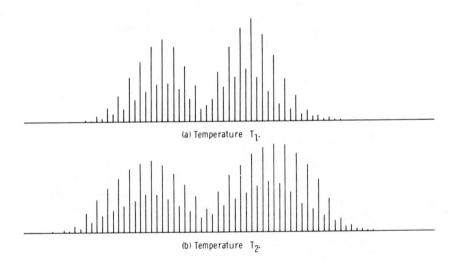

(a) Temperature T_1.

(b) Temperature T_2.

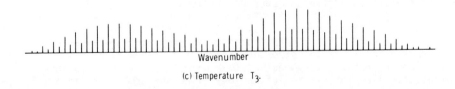

Wavenumber

(c) Temperature T_3.

Figure 1. Temperature variation of line spectrum of single gas system -
           nitrogen. $T_1 \ll T_2 \ll T_3$.

temperature T_1, most of the nitrogen molecules populate the low energy
levels and the line spectrum is "grouped" near the ground state (j = 0).
As the temperature of the gas is increased, more of the molecules populate
the high energy states and the spectrum appears more evenly distributed.
Of course, the spectrum in Figure 1 is highly idealized. A more representative spectrum is shown in the experimental plot in Figure 2 for an

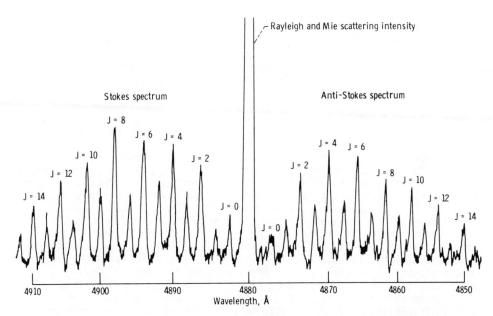

Figure 2.   Analog data plot of rotational Raman spectrum of nitrogen.

exciting laser wavelength of 488.0 nanometers. The same relative peak
heights are evident in Figure 2 but what was in theory an infinitesimally
narrow line appears in the observed spectrum as a broadened peak.

It has been shown in Ref. 3, that for an analysis of a single-
gas-system spectra, it is generally sufficient to consider only the basic
line intensities I_j. Therefore, one can accurately approximate the spectrum in Figure 2 by a line spectrum and only peak heights need to be
measured in experiments of this type. Unfortunately, this approximation
is not satisfactory for analyses of many gas mixtures such as the oxygen-
nitrogen mixture in air because the line positions of the component spectra
are different and partial or complete overlapping is likely. Since the
peak height is generally the experimentally determined parameter in temperature measurement techniques, it is then necessary to take into account
contributions to each peak made by adjacent peaks.

An analysis was performed in Ref. 6 to calculate the spectra
of gas mixtures, with an emphasis on air, while evaluating line broadening
effects such as the Doppler broadening of the laser line, natural line
broadening, and pressure broadening. Also, included in the analysis were
the broadening effects of a typical experimental apparatus in order to

allow exact comparison of a calculated spectrum to its empirical counter-
part. In fact, as was expected, the slit width of the monochromator used
in Ref. 6 was a dominant factor in establishing the shape and bandwidths of the final spectra. As a result of this analysis, very good
agreement was obtained between the calculated and experimental spectra of
air. An example of this agreement is shown in Figure 3 for a monochromator
slit width of 1.84 cm^{-1}.

Throughout the study reported in Ref. 6, particular attention was
directed toward the temperature dependence of an air spectrum and its
application to temperature measurement techniques. The temperature
dependence of the Raman spectra of air is illustrated in Figure 4.
Experiments were conducted over a range of actual air temperatures (i.e.,
from 243 to 313°K) and good agreement was found between the calculated
temperature dependence and experimental data.

One promising method of obtaining temperature measurements from Raman
spectra involves taking the ratio of the intensities of two distinct Raman
lines or two spectral intervals along the Raman spectra. This method offers
some distinct advantages when applied to a remote, real-time temperature
measuring system. Because the two intensities in question can be measured
simultaneously by a number of techniques (e.g., optical interference filters)
high temporal resolution can be obtained and actual real-time measurements
are possible. Also because the rotational Raman lines are contained in a
narrow wavelength or wavenumber region they are, to a good approximation,
affected equally by most extraneous and uncalibrated effects such as the
intervening atmospheric path transmission and the response characteristics
of the detection apparatus. Consequently, when a ratio is taken of intensities distributed along the spectra, these unknown factors are cancelled
and the result depends only on temperature. In practice, the accuracy of
this proposed method of temperature measurement depends on how rapidly
the Raman intensity ratio varies with temperature and how accurately the
ratio can be measured. The method of selecting an optimum intensity ratio
yielding a maximum temperature measurement accuracy can be illustrated
by considering a single gas system and the ratio of two discrete line
intensities. If $\Delta(I_{j_1}/I_{j_2})$ is the error in the measurement of I_{j_1}/I_{j_2}
(where j_1 and j_2 are Raman line number designations) and S is the
variation of the line intensity ratio with respect to temperature, the
error in the measurement of temperature is

$$\Delta T = \frac{\Delta(I_{j_1}/I_{j_2})}{S} \; .$$

The variation of the Raman line intensity ratio with respect to temperature is

$$S = \frac{\partial(I_{j_1}/I_{j_2})}{\partial T} = \frac{K(j_1,j_2)}{T^2} \left(\frac{I_{j_1}}{I_{j_2}}\right) ,$$

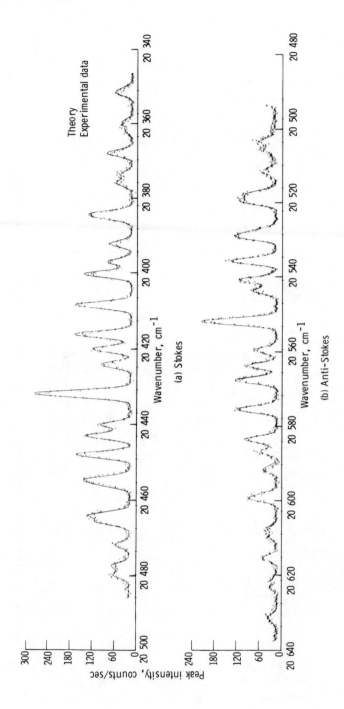

Figure 3.   Comparison of theoretical and experimental spectra of air.

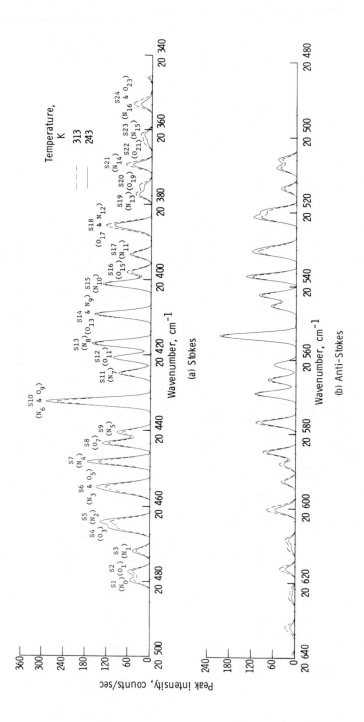

Figure 4.   Comparison of theoretical spectra of air for temperatures of 243 and 313°K.

where $K(j_1,j_2)$ is a known constant for any particular line ratio. Maximizing S by the proper choice of j_1 and j_2 yields the most sensitive measurements over a given temperature range. For Θ/T values of the order of 10^{-2}, S monotonically approaches a maximum with increasing values of j_2 for values of j_1 around 4 to 6. Unfortunately, the Raman line intensities rapidly decrease at large values of j, where the exponential term in the intensity equation dominates. Increasing S ultimately causes the accuracy of the temperature measurement to decrease because of the difficulties in accurately measuring the low level intensities of the high j-value lines. The optimum selection of the Raman line is, therefore, primarily determined by an analysis of the line measurement error,which can be predicted. The minimum measurement error can be predicted by assuming that the bulk of the line intensity measurement error is due to the inherent scattering error which follows a standard Poisson distribution approximation. Detailed calculations of this type are contained in Ref. 3 along with comparisons of the results with experimental data.

Similar error calculations were made for temperature measurements of air in Ref. 6. In all cases the data obtained in the experiments

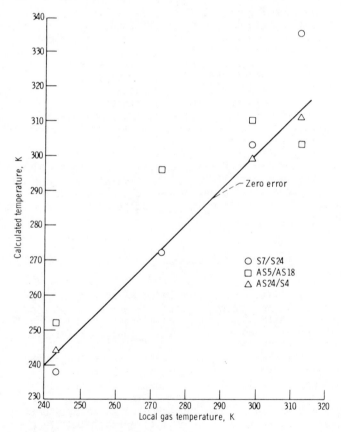

Figure 5.   Temperature measurement results calculated from the Stokes line ratio S7/S24, the anti-Stokes line ratio AS5/AS18, and the anti-Stokes/Stokes line ratio AS24/S4.

yielded temperature measurement accuracies in agreement with those predicted. In the case of air and the Raman apparatus employed for that
study, the predicted accuracy was approximately ±20°K for a single intensity ratio. Some temperature measurement data are shown in Figure 5
where the Raman line designations are as indicated in the spectrum of
Figure 4.

Although this temperature measurement error may seem high, it is
encouraging that it is near that predicted for the particular experimental apparatus in question. Because of this, it is not unreasonable to
predict considerably greater accuracies through improved system design
and the use of other ratio techniques such as using the average of
several ratios or using the ratio of spectral bands rather than discrete
spectral lines. In fact, in the experiments of Ref. 3, the temperature
measurement accuracy was shown to be increased by at least twofold
by employing ratios of summed intensities of adjacent spectral lines.
These tests were considered to approximate the results that would be
obtained with optical interference filters. The concept of employing
optical interference filters in this temperature measurement technique
and further applying the overall technique as an actual remote sensing
measurement tool has been demonstrated to be feasible. Atmospheric
temperature measurements have been made by utilizing the Raman shifted
return of a lidar, or optical radar, system. These measurements were
made along a horizontal path at temperatures between -20°C and +30°C
and at ranges of about 100 meters. From ten laser shots, or one minute
of data acquisition, absolute temperature measurement accuracies of ±3°C
have been obtained with a range resolution of about 5 meters. Again, this
measurement accuracy compares well with that predicted for this particular
lidar unit and, consequently, it is suggested that a field-application
version of this system could be built with significant improvements in
both absolute accuracy and range.

The potential of utilizing Raman scattering as a remote temperature
sensing technique has been established but considerable study and testing
is still necessary before this potential can be exploited to the point
that it results in usage on a large scale, routine basis.

REFERENCES

1. John Cooney, "Satellite Observation Using Raman Component of
    Laser Backscatter," in Proceedings of the Symposium of Electro-
    magnetic Sensing of the Earth from Satellites, ed. by Ralph Zirkind,
    (Polytechnic Institute of Brooklyn Press, 1967), pp. P1-P10.

2. J. Pressman, C. Schuler and J. Wentink, "Study of Theory and
    Applicability of Laser Technique for Measuring Atmospheric
    Parameters," Rep. GCA-TR-68-6-N, GCA Corp.; also NASA CR-86134 (1968).

3. Jack A. Salzman, William J. Masica, and Thom A. Coney, "Determination
    of Gas Temperatures from Laser-Raman Scattering," NASA TN D-6336 (1971).

4. R. S. Hickman and L. H. Liang, Rev. Sci. Instr. 43, 796 (1972).

5.   M. Lapp, C. M. Penney, and J. A. Asher, "Application of Light-
      Scattering Techniques for Measurements of Density, Temperature, and
      Velocity in Gasdynamics," Aerospace Research Laboratories Report No. ARL
      73-0045 (1973).

6.   Thom A. Coney and Jack A. Salzman, "Determination of the Temperature
      of Gas Mixtures by Using Laser Raman Scattering," NASA TN D-7126
      (1973).

                          SALZMAN   DISCUSSION

     BERSHADER - Would you mention how you obtained the measurements
shown on the last slide (i.e. Figure 5)?

     SALZMAN - The data were taken in the laboratory, not outdoors.
The data taken with our outdoor lidar unit have not been presented as
yet but will be presented at the Laser Radar Conference to be held in
Williamsburg during June. The outdoor measurements were made along
a horizontal path utilizing a control volume of air at a known
temperature.

     HENDRA - One of the fundamental problems, it seems to me, with the
system you describe is that you are focusing a laser out into space and
looking for a circular image coming back. On the other hand, spectrometers
like a slit image. This problem was faced by far infrared spectroscopists
many years ago, and their solution was to go to the interferometer, which
has a round entrance pupil. Have you thought of that?

     SALZMAN - Currently, we use optical interference filters which have
circular apertures. Only the laboratory measurements were made using a
monochromator and continuous-wave Argon ion laser. For the outdoor
measurements we used two interference filters. One of these filters
passed a band of line intensities in the far portion of the rotational
anti-Stokes spectrum of air while the other filter passed a band in the
near portion of the anti-Stokes spectrum. The ratio of the intensities
passed through these filters is a direct indicator of the air temperature.

     SMITH - Just a comment here. I'm sure you are aware of it, but for
anyone getting into rotational Raman temperature measurements for the first
time, it is important to emphasize that one pick the lines, or groups of
lines, to be compared with care. For example if lines of equal J on the
Stokes and anti-Stokes wings are compared, their ratio is very insensitive
to temperature. In this case, when the log term is calculated, the
temperature expression blows up numerically.

     SALZMAN - Yes, in those cases where the complement in the log term
is very near unity, the calculation blows up. This is one of the things
which must be considered when choosing the lines or groups of lines to
be used in the ratio because this can occur even if the ratio of the

signal intensities is not near unity. In general, you don't want the ratio to be very large because of the limited dynamic range of most instruments. There are optimum ratio combinations which can be calculated for a particular measuring system and temperature range.

SESSION III
RESONANCE EFFECTS; REMOTE PROBES;
EXPERIMENTAL ADVANCES
Session Chairman: W. Kiefer

LIGHT SCATTERING AND FLUORESCENCE IN THE APPROACH TO
RESONANCE - STRONGER PROBING PROCESSES*

by

C. M. Penney

General Electric Corporate Research & Development
Schenectady, N. Y.

ABSTRACT

For some time there has been hope that many of the desirable
qualities of Raman scattering for probing gas systems can be retained
with orders of magnitude intensity enhancement by tuning the incident
light near or into absorption regions. In this paper we present some of
the results along this line obtained at GE and elsewhere. Calculations
indicate that strong enhancement of electronic Raman scattering may be
observed for many monatomic species upon excitation in near resonance;
there is a good chance that similar results will be obtained for free
radicals such as OH in flames. Moderately strong scattering has been
observed from halogen molecules excited in absorption regions above
their dissociation limits in the visible; it is likely that even stronger
scattering will be observed from ozone through this process. Strong
fluorescence from NO_2, SO_2 and I_2 (below the dissociation limit) can
be observed even at atmospheric pressure. The probe-relevant
characteristics of these re-emission processes are discussed along
with some potential applications.

I. INTRODUCTION

Advantages of Raman scattering for gas probe applications were
described in the introduction to this Proceedings. In particular, Raman
scattering is instantaneous, and insensitive to the quenching effects
that characterize fluorescence, and make it difficult to use as a quantitative probe in many cases. However ordinary non-resonance Raman
scattering is an extremely weak process. Because of this weakness,
Raman probes often encounter limitation of expense, size, weight, power
consumption and/or safety that unduly limit their performance range.
Thus there is motivation to search for stronger processes that preserve

---

*This work was supported in part by the NASA-Langley Research Center.

191

to a useful degree the desirable probe characteristics of Raman scattering. The purpose of this talk is to describe results we and some
others have obtained that are relevant to this search. We shall limit
ourselves to techniques involving incident light in the visible and
near UV, because that is the regime in which we have direct experience,
and have found some definite promise.

II.   THEORETICAL FOUNDATION

The possibility for strongly enhanced Raman scattering excited
near resonance has been recognized for many years. The potential
enhancement is evident from the resonance denominators in the expression
for the Raman scattering differential cross section (cross section per
unit solid angle). The expression can be put into the form

$$\left(\frac{d\sigma}{d\Omega}\right)_{12} = \frac{\omega_2^4}{c^4\hbar^2} \sum_n F_n \sum_f{}' \left| \sum_g{}' \left\{ \frac{(D_2)_{fg}(D_1)_{gn}}{\omega_{gn} - \omega_1} + \frac{(D_1)_{fg}(D_2)_{gn}}{\omega_{gf} + \omega_1} \right\} \right|^2 \qquad (1)$$

Here ω_1 and ω_2 are the angular frequencies of incident and scattered light.
Thus $\omega_2^1 = 2\pi c/\lambda_2$, where λ_2 is the scattered light wavelength. The sub-
scripts n, g and f designate initial, intermediate and final quantum
states of the molecule. The symbols D_1 and D_2 represent components of
the dipole moment in the directions of electric field polarization of the
incident and observed scattered light, and ω_{gn}, for example, is the
angular frequency corresponding to the difference between energies E_g
and E_n of an intermediate and initial state; i.e., $\hbar\omega_{gn} = E_g - E_n$. The
fraction of molecules in the initial state n is represented by F_n. The
prime on the sum over final states indicates that, for each initial state,
this sum must be restricted to only those final states which contribute
to the observed spectral line (i.e., for which $\omega_2 = \omega_1 - \omega_{fn}$ lies within
the observed spectral range). The sum over intermediate states g is
meant to include an integral over the continuum of positive energy
states (dissociative and ionized states).

Although accurate Raman scattering cross sections can be calculated
for H_2 and D_2 starting from Eq. (1), such calculations have not met with
success for heavier molecules. Furthermore, in the general case it has
not been possible to calculate Raman scattering cross sections from
experimental measures of the magnitude of dipole moment matrix elements,
such as oscillator strengths. There are two difficulties with this
calculation: First, the magnitude of all potentially important contributions (including continuum contributions) must be known. Second, the
sum over intermediate states involves terms that can have different
signs. (In the general case, these terms can be complex with different

phases.) Since these terms are combined <u>within</u> the **absolute** square, it is possible that interferences can occur between them. Thus the sign (or phase) of the dipole matrix element is required in addition to its magnitude for cross section calculations. In general, only for coherent Rayleigh scattering (f = n) does Eq. (1) simplify sufficiently to allow a direct correlation to another, more easily measured physical quantity, the refractive index.[2] For other types of scattering, the utility of Eq. (1) has been primarily limited to determination of selection rules and the angular dependence of the scattering. However, close to resonance with a particular absorption line, this equation can be simplified to the point where quantitative calculations are accessible.

III. NEAR RESONANCE SCATTERING

When the exciting line is sufficiently close to resonance with an isolated absorption line, one transition from a particular initial level a to an intermediate level r can give the predominant contribution. In this case the cross section becomes

$$\left(\frac{d\sigma_{12}}{d\Omega}\right) = \frac{\omega_2^4}{c^4 \hbar^2} F_a \frac{|(D_2)_{fr}|^2 \, |(D_1)_{ra}|^2}{(\omega_{ra} - \omega_1)^2 + \gamma_r^2/4} \tag{2}$$

Here we have assumed that the resonance occurs in a term of the first type within the absolute square in Eq. (1) because, as a result of energy conservation, resonances with terms of the second type can only occur in highly excited systems where the intermediate state lies below the initial and final state.[3] The natural damping constant γ_r has been inserted and we have assumed for simplicity that the transition is associated with single initial and final states. Expressing the absolute squares of matrix elements in terms of oscillator strengths, we can write Eq. (2) in the form

$$\left(\frac{d\sigma_{12}}{d\Omega}\right) = 2\pi F_a \frac{g_f}{g_r} \left(\frac{\omega_2}{\omega_1}\right)^3 \left(\frac{e^2}{mc^2}\right)^2 \left\{\frac{\omega_1^2}{(\omega_{ra} - \omega_1)^2 + \gamma_r^2/4}\right\} f_{rf} f_{ra} \tag{3}$$

where we have taken into account the degeneracy of molecular states by including the multiplicaties g_r and g_f of intermediate and final states. Here f_{ra} and f_{rf} are absorption oscillator strengths, and the squared electron radius is given by

$$\left(\frac{e^2}{mc^2}\right)^2 = 7.94 \times 10^{-26} \text{ cm}^2 .$$

Equation (3) can be used with experimental or theoretical values of oscillator strengths to calculate the cross section as a function of the separation from resonance $\omega_{ra} - \omega_1$. There is an interesting controversy associated with this calculation; namely, does the resulting cross section describe scattering or fluorescence?[4,5] The controversy extends to the definitions of these two terms.[6] For the purposes of this paper we shall assume that fluorescence is a process that displays a measurable time decay at low gas pressure, and is quenched significantly by intermolecular collisions at high gas pressure (typically $\gtrsim$ 1 torr). On the other hand, scattering is effectively instantaneous and does not suffer quenching;[7] i.e., its intensity per molecule is insensitive to gas constituency and pressure (at least up to several atmospheres).[8]

The basic question regarding the transition from scattering to fluorescence is illustrated pictorially in Figure 1. When the excitation

A. NON - RESONANCE

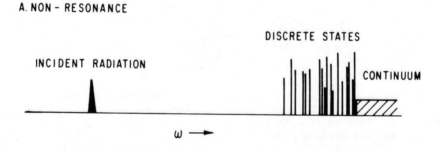

B. APPROACH TO RESONANCE

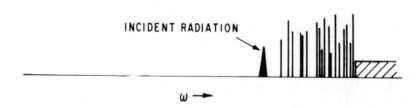

Figure 1.   Illustration of nonresonance excitation (A) and approach to resonance (B).

is far from resonance, as shown in Figure 1A, the re-emission is known to display the properties of scattering. As resonance with a particular line is approached, as in Figure 1B, the properties must begin to change at some point to those of fluorescence. The question becomes then - how does this change progress as the separation from resonance decreases? At least at low pressure, whenever the separation from resonance becomes small enough so that the contribution from a single line predominates, it is likely that the polarization and angle-dependent characteristics of the re-emission will become that of the corresponding fluorescence, because these properties are then determined by the same matrix elements in both cases. But what about the time-dependence and quenching properties that characterize scattering as we have defined it? Our preliminary theoretical analyses (concerning primarily the time-dependence) indicate that, in a gas sufficiently rarified so that interactions between molecules are not important, Eqs. (2) and (3) describe scattering (e.g. an instantaneous process) at separations from resonance that are large in comparison to γ_r and the Doppler broadening width Γ_D, and describe fluorescence on resonance. In more dense gases, homogeneous and inhomogeneous line broadening and quenching introduce complications regarding the time dependence and pressure sensitivity of the cross section that have been resolved only partially by theory and experiment[3,4] to date. Nevertheless, it appears reasonable to use Eq. (3) to obtain initial estimates of near resonance re-emission that might display the favorable characteristics of Raman scattering under conditions of interest.

A. Near Resonance Scattering from Atoms

Equation (3) leads to encouraging predictions of large cross sections in cases involving near resonance with a strong, isolated atomic line. For example, in Figure 2 we show calculated cross sections for electronic Raman scattering from aluminum atoms,[2] expressed as functions of wavelength. The Raman shift for these transitions is about 20 Å. An enhancement by 10^5 over ordinary Raman scattering cross sections is predicted at separation from resonance on the order of 50 Å. Similar predictions obtain for many other atoms (e.g. Ga, B, Tl, Cu, rare earths, Fe, In, Ni, C, etc.) whose energy levels and selection rules make possible a strong electronic Raman transition. Experiments to verify the characteristics of this type of scattering appear feasible now, because of the rapid development of tunable lasers. It may be that such experiments will reveal a sensitivity to background gas more characteristic of fluorescence than scattering. If so, then perhaps we should call the process off-resonance fluorescence, rather than near resonance scattering. But if background sensitivity is encountered at such large separations from resonance, it should provide fascinating information about line broadening in the far wings of a line, beyond the region observable in absorption. Thus, regarding re-emission excited near resonance with an isolated atomic line, we see the possibility of strong scattering useful for probing applications, and alternatively, an opportunity to study broadening phenomena in the far wings of the line.

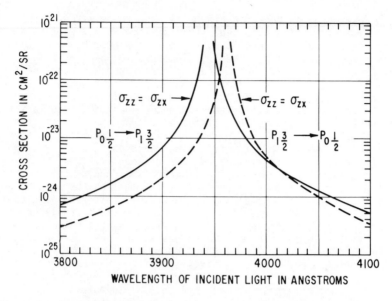

Figure 2.   Calculated electronic Raman scattering cross sections
            for aluminum atoms (See **Ref.** 2).

B. Near Resonance Scattering from Molecules

For molecules, the situation is less favorable because of the usual
multitude of vibrational and rotational transitions, which dilute the
electronic transition.   This phenomenon reduces the potential enhancement
in three ways:

1)   The fraction of molecules in the particular initial level that
     leads to near resonance transitions can be much smaller than 1.
     If the molecules are effectively all in their vibrational
     ground state, this fraction is equal to the population fraction
     of the initial rotational level.

2)   Average individual molecular line oscillator strengths are
     smaller than corresponding atomic oscillator strengths by a
     factor on the order of the number of significant non-degenerate
     upward transitions from the initial state to states within the
     rotational-vibrational manifold of the excited electronic state.

3) A substantial number of significant downward transitions can proceed from the intermediate level in near resonance, such that the re-emission is divided among this number of lines.

The reduction in enhancement through these effects can easily be several orders of magnitude or more in typical cases. Consequently, in order to obtain significant enhancement in near resonance to a molecular line, we expect that it will be necessary to be much closer to the line than in the previously described case involving a strong atomic transition. This expectation is strengthened by the fact that many molecular spectra consist of very closely spaced lines. Quantitative measurements obtained from excitation very close to a resonance line require precise control over laser frequency. Furthermore, they face increased possibility that collisions will introduce fluorescence-like quenching effects. Nevertheless, some promising results have been obtained. Fouche and Chang[9] and St. Peters et al[5] have reported strong re-emission from iodine vapor (I_2) excited very near resonance (within 0.01 nm) with individual molecular lines. In N_2 near STP, shifted line re-emission from I_2 has been observed that is about six orders of magnitude stronger, per molecule, than vibrational Raman scattering from N_2. Furthermore, this emission shows a quenching dependence on background gas that is much weaker than that usually observed for fluorescence. This observation has been explained as a near balance between the opposing effects of collisional broadening (which increases absorption in the wings of a line) and quenching (which reduces the probability of re-emission following absorption).[4,5] The increase in absorption is illustrated in Figure 3. The resulting

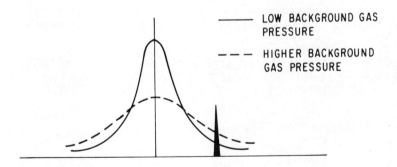

Figure 3.   Illustration of line broadening and resulting increased absorption on line wings.

variation of re-emission per molecule in an actual case is shown in
Figure 4. In this particular case the re-emission rises initially with

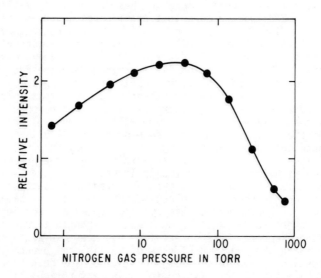

Figure 4.   First Stokes band signal intensity from monochromatically
             excited I_2 vapor vs. pressure of N_2 gas added to the cell.
             The signal is relative to the signal with no N_2 added.
             The spectrometer slit width is 34 cm^{-1}. The incident
             wavelength is 1750 - 1800 MHz from the center of the
             strong resonance within the bandwidth of the 514.5 nm
             argon ion laser line.

background gas pressure, and then reaches a peak. Near the peak, the
re-emission varies by less than 10% as the pressure changes through a
factor of ten.

Following, in essence, the analysis presented in Refs. 4 and 5,
the variation of I_2 re-emission with nitrogen gas pressure can be
explained by expressing the cross section for re-emission in the form

$$\left(\frac{d\sigma}{d\Omega}\right)_{re-em} \approx \frac{\sigma_{abs}}{4\pi} P\phi \qquad (4)$$

Here P is the probability of decay of the excited state by re-emission,
and ϕ is the fraction of the re-emission that goes into the observed
spectral channel. The absorption cross section is given by

$$\sigma_{abs} = F_a \left\{ \frac{\pi^2 c^2 \gamma_{ra}}{\omega_{ra}^2} \right\} \left\{ \frac{\Gamma/2\pi}{(\omega_1 - \omega_{ra})^2 + \Gamma^2/4} \right\} \tag{5}$$

where, in addition to quantities defined previously, γ_{ra} is the transition probability (per unit time) for the resonance line, Γ is the total line width, given by

$$\Gamma = \gamma_r + K_1 P_1 + K_2 P_2 + \cdots \tag{6}$$

and γ_r is the probability per unit time of decay by radiation. (Note γ_{ra} is one component of γ_r, which is the sum of such components over all possible downward transitions.) We have assumed that the collision broadening is proportional to the partial pressures P_1, P_2, $\cdots$ of the various gas constituents. The effects of inhomogeneous broadening (Doppler broadening, interaction line shifts, etc.) have been neglected for the moment. The probability of decay by re-emission is taken to be

$$P = \frac{\gamma_r}{\Gamma - \Gamma_e} \tag{7}$$

where Γ_e is the contribution to broadening by "elastic" collisions, i.e., collisions that do not degrade the excitation beyond the observed spectral channel.

From Eqs. (4), (5), (6) and (7), we obtain

$$\left(\frac{d\sigma}{d\Omega} \right)_{re-em} = F_a \phi \left\{ \frac{\pi^2 c^2 \gamma_{ra}}{\omega_{ra}^2} \right\} \left\{ \frac{\Gamma/2\pi}{(\omega_1 - \omega_{ra})^2 + \Gamma^2/4} \right\} \left\{ \frac{\gamma_r}{\Gamma - \Gamma_e} \right\} \tag{8}$$

which can be put into the simplified form

$$\left(\frac{d\sigma}{d\Omega} \right)_{re-em} = \left\{ \frac{(constant)}{(\omega_1 - \omega_{ra})^2 + \Gamma^2/4} \right\} \left\{ \frac{\Gamma}{\Gamma - \Gamma_e} \right\} \tag{9}$$

Here, the "constant" contains factors that do not change with pressure.

Equations (8) and (9) can be shown to be equivalent to Eq. (2) or (3) for gas systems at pressures low enough so that collision effects are negligible. However, within our simple model, Eqs. (8) and (9) also apply when collisions are significant. They lead to several interesting predictions. In particular, if $\Gamma_e << \Gamma$, then the second term of Eq. (9)

is near unity. Then, for separations from resonance such that
$(\omega_1 - \omega_{ra})^2 >> \Gamma^2$ over the pressure range of interest, the re-emission per
molecule will be independent of gas pressure and constituency; that is,
the re-emission will display a scattering-like insensitivity to quenching.

At first glance, this result would seem to answer our previous
question about the distinction between scattering and fluorescence.
However the model is too simple to give a distinction in which we can
be confident. The conclusions it leads to are not altered by inclusion
of Doppler broadening provided that the separation from resonance is much
larger than the Doppler broadening width Γ_D, because the Gaussian
distribution of Doppler broadening falls off very rapidly in the wings.
However, other types of inhomogeneous broadening (including, in particular, satellite band effects in the wings[10]) could introduce significantly
different affects.

In the intermediate cases described in Refs. 4 and 5, where
$(\omega_{ra} - \omega)^2$ is only somewhat larger than Γ or Γ_D, the pressure dependence
of the ratio $\Gamma/(\Gamma-\Gamma_e)$ can be invoked to explain the initial increase of
the scattering with increasing background gas pressures, and to explain
the subsequent peak, whose position depends on the separation from
resonance.   The experiments involve excitation by argon and krypton
lasers that are single-moded and tuned within a gain curve by a
tilted etalon. The strong observed enhancement opens the possibility,
for example, for observation of trace concentrations of I_2 vapor in
the atmosphere, where it might serve as an indicator of marine biological
activity.[11]

In addition to I_2, there are some other molecules that present
accessible strong absorption lines. Notable among these are Br_2, and
radicals found in flames and gas discharges, such as OH. Unfortunately,
this group encompasses only a small fraction of the molecules of interest.
Furthermore, it is not obvious that the advantage of reduced sensitivity
to quenching gained by off-resonance excitation is greater than the
disadvantage of reduced emission incurred thereby. In fact, this trade-
off appears to depend sensitively on the characteristics of the light
source and measurement environment in each particular case. Thus we
are led to look beyond near-resonance excitation of an isolated line
for other phenomena that can produce quantifiable strong re-emission
from molecules.

C. Near Resonance to a Molecular Band

A number of molecules display absorption in isolated bands, such as
those of NO illustrated in Figure 5. Thus it is of interest to estimate
the enhancement to re-emission contributed by the near proximity of one
of these bands. A precise calculation of this contribution requires
attention to the affects of interference between contributions to the
absolute square in Eq. (1), as mentioned previously. However, at a
separation from resonance that is large compared to the width of the
band, it can be shown that possible interferences in the band contribution tend to be small, such that the contribution of the band to the

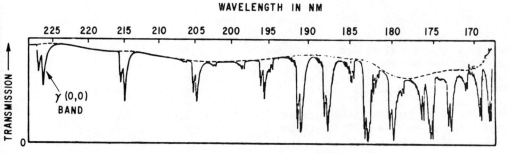

Figure 5. Nitric oxide absorption bands in near UV.

re-emission is approximately proportional to its overall strength.
Within this approximation, we obtain from Eq. (2)

$$\left(\frac{d\sigma}{d\Omega}\right)_{12} = 2\pi \left(\frac{e^2}{mc^2}\right)^2 \left(\frac{\nu_2}{\nu_1}\right)^3 \left(\frac{\nu_1}{\nu_1-\nu_R}\right)^2 f^2 \sum_{n,r} P_n \, |<r|n>|^2 \, |<r|f>|^2 \qquad (10)$$

This result is similar to that obtained previously by others.[3,12] Here f
is the electronic absorption oscillator strength, P_n is the fractional
population of the initial vibrational state R, and $|<r|f>|^2$, for
example, is the Franck-Condon factor between vibrational states r and
f, normalized such that

$$\sum_r \, |<r|f>|^2 = \sum_f \, |<r|f>|^2 = 1$$

The effective resonance frequency is denoted simply by ν_R.

   Using Eq. (9), we can estimate the contribution of the $\gamma(0,0)$ band
of NO, centered at 226.5 nm, to the re-emission excited at 230 nm. The
electronic oscillator strength and Franck-Condon factor $|<0|0>|^2$ have
been measured by Bethke,[13] who obtained f = .0025 and $|<0|0>|^2$ = 0.165.
We will assume that $|<f|o>|^2$ = 0.1, where f is the largest Franck-Condon
factor for a downward transition producing a shifted re-emission line
analogous to vibrational Raman scattering. These values yield

$$(\frac{d\sigma}{d\Omega})_{12} \approx 2 \times 10^{-28} \ cm^2/sr,$$

which is about 40 times larger than the cross section predicted by
an extrapolation from measurements in the visible at 514.5 nm using
the usual $(\omega_2)^4$ dependence. Thus Eq. (9) leads to a prediction of
modest enhancement for this case. Of course the predicted enhancement
would be much larger if the electronic transition were strong. Note
that the cross section is proportional to the square of the electronic
oscillator strength and this, for the transition considered, is about
1/400 of that for a strong transition $(f \sim 1)$. Thus, although the
calculation just presented does not indicate much promise for strongly
enhanced scattering from NO, it does suggest that such enhancement
might be obtained in near resonance to a <u>strong</u> compact band. In
such a case, the allowable separation from resonance might be much
larger than in the case of near resonance to a single molecular
line, relaxing requirements on laser wavelength stability.

IV. <u>RESONANCE SCATTERING IN A DISSOCIATIVE BAND</u>

So far we have considered excitation near resonance, but
sufficiently far removed from it so that the re-emission has some
chance of displaying scattering-like characteristics; i.e., independence
from quenching. However, in 1970, Holzer, Murphy and Bernstein[7] identified a re-emission process with scattering-like properties resulting
from on-resonance excitation. This process occurs when the excitation
is within a dissociative band. Following such excitation, the
molecule must either undergo radiative de-excitation in a very short
time, or fly apart. In this case, there is little time for quenching
collisions to occur, and the re-emission consequently displays
scattering-like insensitivity to quenching. Holzer et al observed
this process by using an Ar+ laser to irradiate the halogen vapors I_2,
Br_2, Cl_2, IBr, etc., all of which have dissociation limits near 500 nm.
In particular, they observed that the strongest scattering line from
I_2 vapor excited at 488 nm is about 1000 times stronger, per molecule,
than vibrational Raman scattering from N_2 at the same wavelength. Thus,
this type of re-emission from I_2 retains the desirable probe characteristics of Raman scattering in combination with substantial enhancement.

It is quite likely that even more encouraging results can be
obtained when this type of re-emission is observed from other molecules.
For example, the very strong Hartley absorption band in O_3,[14] centered
at 250 nm, is a dissociative band. This band is broad (about 50 nm
half width) and displays little structure under high resolution. We
can obtain a very rough estimate of its re-emission cross section from
the following argument: The absorption coefficient at the peak of the
Hartley band at STP is about 300 cm^{-1} atm^{-1}, which corresponds to
$\sigma_{abs} \approx 10^{-17} \ cm^2$. The re-emission differential cross section can be

estimated from an approach similar to the one used previously. Consider

$$\left(\frac{d\sigma}{d\Omega}\right)_{em} \simeq \frac{1}{4\pi} \sigma_{abs} P\phi \qquad (11)$$

where again P is the probability of de-excitation by re-emission and ϕ is the fraction of this re-emission going into the observed line. If the primary de-excitation is dissociation, then

$$P = \frac{\gamma_r}{\gamma_r + \gamma_{dis}} \qquad (12)$$

where γ_r is the rate of de-excitation by radiation, and γ_{dis}, by dissociation. (We assume that γ_{dis} is much greater than collision de-excitation rates.) From Figure 6, the integrated absorption coefficient for the Hartley band is

$$K \equiv \int k_2 d\lambda \simeq 1.2 \times 10^{-4}$$

from which

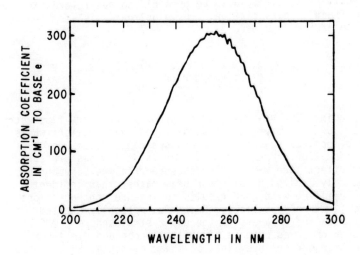

Figure 6.   Ozone Hartley absorption band in near UV.

$$\gamma_r \simeq \frac{8\pi c}{\lambda_o^4 N_o} K \simeq 10^8 \ sec^{-1}$$

Here λ_o is the center wavelength (250 nm) and N_o is the molecule number density. We can estimate

$$\gamma_{dis} \simeq \frac{V}{d} \tag{13}$$

where V is the average velocity of the oxygen atom that is ejected, and d is the distance beyond which dissociation is completed. If the kinetic energy associated with the motion of dissociation is 1 ev, then $V \simeq 4 \times 10^5$ cm/**sec**. We further estimate $d = 4 \times 10^{-8}$ cm and $\phi = 0.1$. Then

$$\frac{d\sigma}{d\Omega} \simeq 10^{-24} \ cm^2/sr$$

This value of cross section is about five orders of magnitude larger than the N_2 vibrational Raman cross section extrapolated to 250 nm by use of the factor ω_2^4. Such a large enhancement, if corroborated by experimental measurements, would make possible remode LIDAR observations of ambient O_3 concentration (~10 ppb) in the lower atmosphere, and, perhaps spatially resolved observations of the stratospheric O_3 level (10-30 km altitude) from a mountaintop observatory (using a slightly longer wavelength (~300 nm) that will penetrate this level). In addition to observations of this type of Raman scattering, it may also be possible to see re-emission from the oxygen atom freed by dissociation, providing that this atom leaves in an appropriate excited state.

V.   FLUORESCENCE

Up to this point, I have avoided discussion of ordinary fluorescence as a gas probe. However, there are several types of situations where this strongest of optical re-emission processes can provide useful quantitative information. One such situation is at low gas pressure, where quenching effects are weak. Fluorescence from the Schumann-Runge bands of O_2, usually excited by electron beam rather than incident light, has been utilized with good results to measure concentrations in low pressure ($\lesssim 1$ torr) gas flows. Recently, optically excited fluorescence from Na and K layers in the upper stratosphere (~100 km altitude) has been observed using a LIDAR-type probe and dye lasers tuned on to the strong resonance lines of these elements.[15,16] These observations have provided quantitative measurements of the spatially-resolved densities of these layers and their variations in, for example, a meteor shower.[17]

Fluorescence can also provide quantitative information in gas near atmospheric pressure. In particular, if the primary quenching effect is due to constituents whose partial pressures are known, then this effect can be determined in controlled experiments, allowing determination of unknown densities of fluorescing species from measurements of the fluorescence intensity.

Such a situation may occur for several interesting pollutant species in the atmosphere. For example, Birnbaum and co-workers[18] have shown that fluorescence from NO_2 excited by argon laser wavelengths in the visible region is sufficiently strong to allow observation of source concentrations and high ambient levels of this gas in the atmosphere. At STP this fluorescence, integrated over its broad spectral distribution, appears to be 10^3 to 10^4 times stronger than N_2 vibrational Raman scattering.[4] If the quenching of this fluorescence is due predominantly to the major constituents N_2 and O_2, whose concentrations are known, then observation of fluorescence intensity (relative, say, to N_2 Raman scattering) will lead to quantitative measurements of NO_2. However, it remains to be shown whether or not H_2O, CO_2 and/or temperature variations will affect the re-emission.

Another example of present interest is provided by SO_2 in the atmosphere, excited by incident light near 300 nm. In our laboratory, we have studied the properties of this fluorescence using a narrow-line (0.005 nm), short pulse (5 nsec), doubled dye laser. Block diagrams of the optical and electronic systems are shown in Figures 7 and 8. The dye laser is excited by a pulsed N_2 laser (AVCO model C950). Tuning is accomplished by a telescope/grating combination similar to that described by Hänsch.[19] The monochromator (Spex model 1800) provides spectral resolution of the re-emitted radiation to about 0.02 nm resolution. The photomultiplier (EMI 9635) pulse train is transmitted to a constant fraction timing discriminator (Ortec model 463) whose output pulses are gated by the time-to-height converter and single channel analyzer (Ortec models 437A and 406A, respectively). The re-emission time distribution can be displayed using the multi-channel analyzer. Observed time resolution of this system is 5 nsec.

A low-resolution absorption spectrum of SO_2 is shown in Figure 9. In our experiments we tuned through portions of the J, H, G and F bands. A region of particularly strong fluorescence was observed in the G band near 299.98 nm (air wavelength, ± .05 nm). The spectral distribution of this fluorescence is shown in Figure 10. Notice the strong re-emission into the ν_1 line.

When N_2 is admitted into the scattering cell, producing a total pressure near one atmosphere, the intensity per molecule of this line is reduced substantially, but it still stands well above the underlying fluorescence intensity, as shown in Figure 11. The intensity variation of this line against background N_2 pressure is shown in Figure 12, and a Stern-Volmer plot of this variation in Figure 13. Finally, we show the variation of the ν_1 line intensity with laser wavelength in Figure 14.

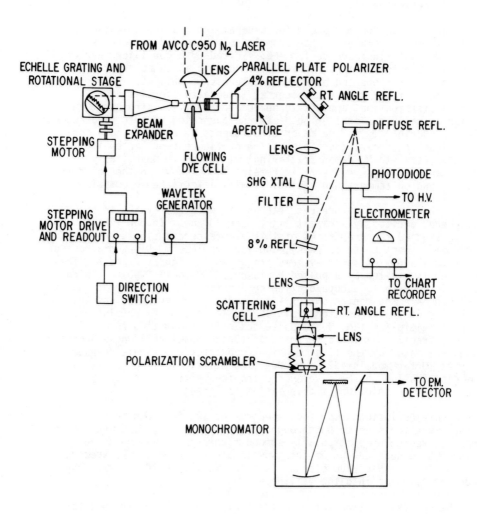

Figure 7.   Schematic of optical system of the doubled tunable
            dye laser.

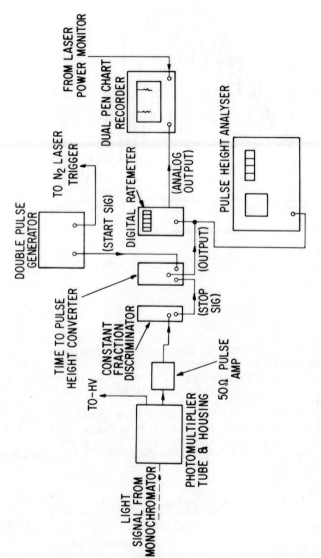

Figure 8.   Schematic of the electronic system used for detector signal analysis with the doubled tunable dye laser.

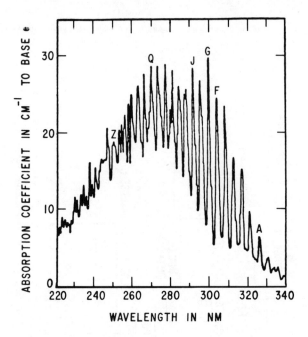

Figure 9.   Low resolution absorption spectrum of SO_2.

We measured the intensity of the SO_2 ν_1-line fluorescence in N_2 by comparison to the N_2 vibrational Raman scattering. The absolute intensity of the latter had been measured previously by comparison to N_2 Rayleigh scattering, whose absolute intensity can be calculated from known values of the refractive index. The result for the fluorescence cross section in N_2 at 700 torr is 10^{-25} cm^2/sr, as shown in Figure 12. This result is believed to be accurate to within a factor of 2. If so, it represents a strong enhancement that is potentially useful for measurements of SO_2 ambient and source concentrations in the atmosphere.

VI.   CONCLUSION

We have discussed processes that produce enhanced re-emission in atoms, and NO, halogen, NO_2, SO_2 and O_3 molecules, among others. Each of these processes retain to some degree the useful properties of Raman scattering for gas probe application. However, it is likely that

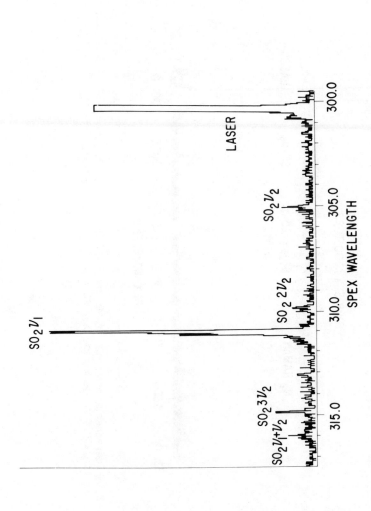

Figure 10. Fluorescence spectrum of low pressure (30 mTorr) SO_2 excited by light at 2999.8 Å (corrected air wavelength) with bandwidth of 0.05 Å.

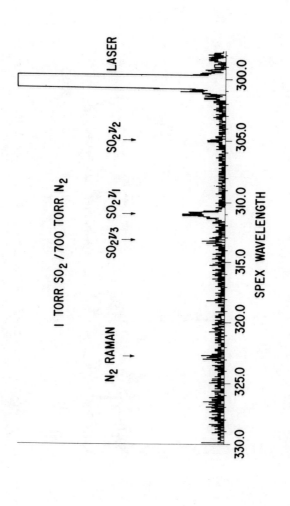

Figure 11. Fluorescence spectrum of SO_2 in 700 torr of N_2. The excitation wavelength is 2999.6 Å (corrected air wavelength) with bandwidth of 0.05 Å.

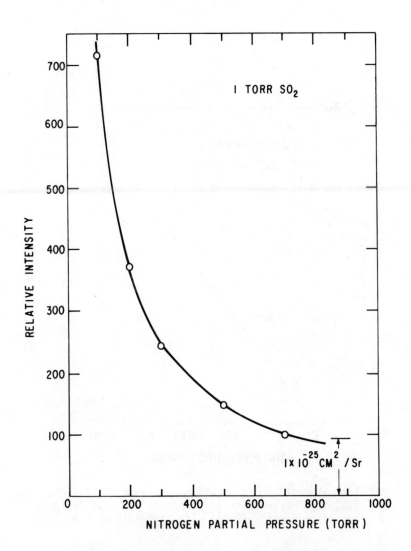

Figure 12.   Intensity of ν_1-line fluorescence from SO_2 as a function
of background air pressure.   The SO_2 fluorescence is excited by incident light at 2999.6 Å with bandwidth 0.05 Å.
The effective cross section for ν_1-fluorescence, measured
by comparison to N_2 vibrational Raman scattering, is also
shown in the figure.

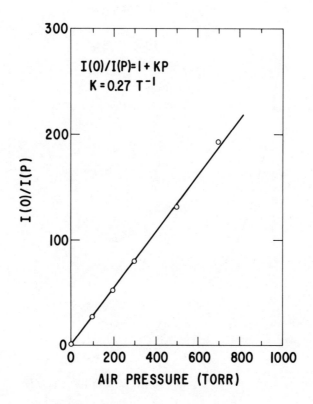

Figure 13.   Stern-Volmer plot of ν_1-fluorescence from SO_2 (Plot of I(o)/I(p) versus nitrogen pressure p). The incident radiation is at 2999.6 Å with bandwidth of 0.05 Å.

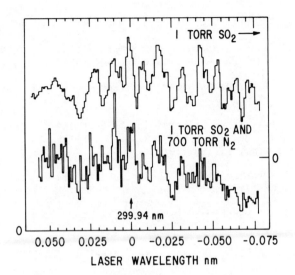

Figure 14. Tuned laser fluorescence spectrum of SO_2 and a SO_2/N_2
            mixture. The spectrometer slit function is trapezoidal
            with a minimum width of 2 Å, centered on the ν_1 line.

some of the desirable qualities of Raman scattering will be
lost when a stronger re-emission process is sought. Precise control
over laser wavelength may be required; it may be more difficult to
relate measured results to desired information; and/or it may be
possible to observe only one or a few species for any particular
incident wavelength. Nevertheless, in difficult probing applications
(long range or very low concentration) the techniques discussed herein
may well provide the best alternative to other techniques or no
information at all.

REFERENCES

1. Near resonance and resonance processes in the IR are discussed
    briefly by Helge Kildal and Robert L. Byer, Proc. IEEE, 59, 1644
    (1971).

2.   C. M. Penney, J. Opt. Soc. Amer. <u>59</u>, 34 (1969).

3.   G. Placzek, "Rayleigh-Streuung und Raman Effekt," in Handbuch der
     Radiologie <u>6</u>, Part 2. Akadamische Verlagsgesellschaft, 1934.

4.   D. G. Fouche, A. Herzenberg and R. K. Chang, J. Appl. Phys. <u>43</u>,
     3846 (1972).

5.   R. L. St. Peters, S. D. Silverstein, M. Lapp and C. M. Penney,
     Phys. Rev. Lett. <u>30</u>, 191 (1973); R. L. St. Peters and S. D. Silver-
     stein, Optics Comm. <u>7</u>, 193 (1973).

6.   Marcel Jacon, Maurice Berjot, Lucien Bernard, C. R. Acad. Sc.
     Paris <u>273B</u>, 956 (1971).

7.   W. Holzer, W. F. Murphy and H. J. Bernstein, J. Chem. Phys. <u>52</u>,
     399 (1970).

8.   G. F. Widhopf and S. Lederman, AIAA Journal <u>9</u>, 309 (1971).

9.   D. G. Fouche and R. K. Chang, Phys. Rev. Lett. <u>29</u>, 536 (1972).

10. See, for example Shang-yi Ch'en and Makoto Takeo, Rev. Mod. Phys.
     <u>29</u>, 20 (1957); W. R. Hindmarsh and Judith M. Farr, J. Phys. B,
     <u>2</u>, 1388 (1969); C. L. Chen and A. V. Phelps, Phys. Rev. <u>A7</u>, 470
     (1973).

11. NASA Langley Research Center Report No. NASA SP-285, p. 76 (1971).
     Copies of this report can be obtained from the National Technical
     Information Service, Springfield, Virginia 22151.

12. Josef Behringer, "Observed Resonance Raman Spectra" in Raman
     Spectroscopy, Herman A. Szymanski, Ed., Plenum Press, 1967.

13. George W. Bethke, J. Chem. Phys. <u>31</u>, 662 (1959).

14. M. Griggs, J. Chem. Phys. <u>49</u>, 857 (1968).

15. M. R. Bowman, A. J. Gibson and M. C. W. Sandford, Nature <u>221</u>,
     456 (1969).

16. F. Felix, W. Keenliside, G. S. Kent and M. C. W. Sandford, "Laser
     Radar Measurements of Atmospheric Potassium," paper presented
     at the Fifth Conference on Laser Radar Studies of the Atmosphere,
     Williamsburg, Va., June 4-6, 1973.

17. M. Jyumonji, T. Kobaysi and H. Inaba, "Measurements of Resonance
     Scattering Cross Sections of Sodium D Lines and Laser Radar
     Detection of the Sodium Layer in the Upper Atmosphere by a
     Tunable Dye Laser," paper presented at the Fifth Conference
     on Laser Radar Studies of the Atmosphere, Williamsburg, Va.,
     June 4-6, 1973.

18.   M. Birnbaum, J. A. Gelbwachs, A. W. Tucker and C. L. Fincher,
      Opto-Electronics, May 1972.

19.   T. W. Hänsch, Appl. Opt. 11, 895 (1972).

PENNEY DISCUSSION

KIEFER - As far as I understood, you do not make any significant distinction between a resonance Raman and a resonance fluorescence spectrum. However, there are some experimental features which clearly say that there must be a difference between both effects. As an example, let's take the depolarization ratio: resonance Raman lines of an X_2 vibration (like the I_2 vibration) are highly polarized, whereas the resonance fluorescence bands are depolarized.

PENNEY - I identified scattering in my talk as an effectively instantaneous process whose intensity per molecule is insensitive to gas pressure. This distinction was chosen both because it is a well established one, and because it conveniently designates those processes useful for gas probing which do not require corrections for background gas pressure. However, this distinction breaks down at very high gas pressures, even for ordinary Raman scattering. It is inapplicable to liquid and solid systems, and even in a gas, it is difficult to establish theoretically.

On the other hand, distinction between scattering and fluorescence by polarization is also quite useful in our work. However, it is not universal: many fluorescence lines are strongly polarized also.

KIEFER - The most striking difference between the resonance Raman and resonance fluorescence effect is that the shifts of the observed lines from the laser line depend only on the property of the electronic ground state of the scattering molecule in the case of resonance Raman scattering, but depend in addition also on the property of the excited electronic state when a fluorescence spectrum is produced. This means that all laser frequencies whose energies are higher than the convergence limit give rise to very similar resonance Raman spectra - especially in terms of exactly the same shifts from the exciting line. Different laser lines whose energies are lower than the convergence limit will - if they hit rotational-vibrational levels of the electronic excited molecule - show a resonance fluorescence spectrum, where the frequency shift of the resulting line can be slightly different for various exciting lines, because now also the spectroscopic constants of the excited electronic state, which are different from those of the ground state, are involved. A resonance Raman spectrum reflects only the property of the electronic ground state of the molecule; a resonance fluorescence spectrum, however, gives information also about the potential function of the electronic excited molecule.

PENNEY - As I understand your comment, it implies that re-emission dominated by transitions through a single intermediate level should be called fluorescence. Several other people have also used this distinction. In fact, it may be the most generally applicable one, but it has the disadvantage of being disconnected in part from the characteristics of the re-emission. For example, consider the re-emission from aluminum atoms discussed in the talk. At an attainable but low aluminum vapor pressure, the cross section calculations presented there imply that the re-emission dominated by a single intermediate level can be seen with excitation several hundred angstroms from resonance. This re-emission should be instantaneous and it may be insensitive to quenching, but even so, the single intermediate level criterion would designate it fluorescence. The point is that the various criteria for distinguishing scattering and fluorescence are each useful, but it seems that they do not overlap in certain special cases. Thus, it may be fruitless to search for a universal distinction.

BERSHADER - Dr. Penney, when you presented those initial formulas, you had expressions which looked very much like the extension of the Lorentz electron theory applied to this problem and those expressions are in fact very similar to the well-known absorption and dispersion formulas of that theory. My question has to do with your assumption of population: you assume that the upper state near which you are scattering has a negligible population compared to the lower state. In the case that it doesn't, or in particular, where you have a population inversion in the case of absorption and refraction, one runs into the interesting phenomena of negative absorption, negative dispersion and so on. My question here is, suppose you have non-negligible population in an upper state and you do this resonance analysis. Does this presumably give you some additional effects which have to be separated out?

PENNEY - Yes, I think so in special cases. If the intermediate state in resonance is in a population inversion, then the scattered light could be amplified. Substantial excitation of high lying electronic levels brings up the possibility of many types of resonance processes, including resonances of the second type in the basic cross section equation - that is, processes where the re-emitted photon is in near resonance rather than the incident photon. However, because Raman scattering provides so much information, none of these possibilities should obviate the use of Raman scattering as a probe in systems with strong electronic excitation.

A population inversion in the initial level of a Raman transition, i.e., a vibrational or rotational level, will produce a commensurately strong hot band or rotational line. This indication provides a straightforward linear diagnostic of non-equilibrium systems.

Finally, stimulated Raman scattering, stimulated Brillouin scattering and gas breakdown are possibilities when highly focussed, energetic, short pulses are used. These processes form fundamental limits in the trade-off between spatial and temporal resolution for Raman probes, as has been noted earlier at this meeting, and of course, their thresholds are lower near resonance. In the SO_2 measurements, we

keep a set of neutral density filters handy to verify that we are
measuring the linear properties.

LEWIS - What are the prospects for using resonance Raman scattering
to make measurements on metastable species?

PENNEY - I think the chances are good. There appear to be accessible
near-resonance transitions from metastables with very large cross sections,
and in fact, some metastable species have already been observed by
Raman scattering. For example, Braünlich and Lambropoulos [Phys. Rev.
Lett. $\underline{25}$, 986 (1970)] saw antistokes Raman scattering from metastable
states of atomic deuterium at a density of just 10^6 cm^{-3}. Very large
cross sections have been calculated for similar transitions from the
$2'S_0$ state of helium atoms by Jacobs and Mizuno [J. Phys. B, $\underline{5}$, 1155 (1972)],
and the unshifted fluorescence from these metastables in a plasma
has been observed by Burrell and Kunze [Phys. Rev. Lett. $\underline{28}$, 1 (1972)].
Among molecules, two likely candidates are nitrogen metastables and
ions, which appear to be present in the upper atmosphere in sufficient
numbers to contribute significantly to the dayglow.

HARVEY - I would like to address this question to Don Leonard.
Why is there such a difference between Inaba's NO_2 cross sections and
yours, for 3371 Å irradiation?

LEONARD - I don't know the answer to this question because I have
never discussed the matter with Prof. Inaba. I believe that in an
oral comment at the laser radar conference in Tuscon in 1972 he claimed
a cross-section for NO_2 10^3 times larger than that for N_2. He also
used the 3371 Å nitrogen laser as a source. My data showed the NO_2
cross-section to be approximately equal to that of N_2.

KIEFER - I would like to make a short comment on the statement
made by Dr. Penney concerning the fact that going into resonance may
increase the Raman signal by orders of magnitude. This is true. However,
the exciting laser beam as well as the back scattered Raman light has
to pass a highly absorbing media, which in fact will decrease the
finally-detectable signal by an appreciable amount. Even so the enhancement by the resonance Raman effect may be two or three orders of magnitude
compared to nonresonance Raman excitation, and the enhancement of the
finally-observed Raman signal may be only a factor of one to ten
depending on the amount of absorbance and the lengths of the light paths.
I expect that the gain due to the resonance Raman effect may be just
compensated by the loss due to the high absorption.

PENNEY - That is a very real possibility, and one that must be
examined for each experimental configuration. It can be examined
conveniently in terms of the quantum efficiency for re-emission of
absorbed photons into the observed spectral line, the quantity represented
by $P\phi$ in my talk. One can show that in many cases, if $P\phi \gtrsim 10^{-6}$, the
absorption loss will be very small in comparison to the resonance enhancement. Furthermore, the situation is even more favorable if one observes
a layer of the absorbing constituent through a relatively nonabsorbing
path, such as SO_2 in a smoke plume or O_3 in the upper atmosphere.

OBSERVATION OF S-BAND HEADS IN THE RESONANCE
RAMAN SPECTRUM OF IODINE VAPOR

by

W. Kiefer

Sektion Physik der Universität München
München, Germany

ABSTRACT

Resonance Raman scattering has been observed from iodine vapor
excited by light above the dissociation limit. The spectra are characterized by a progression of vibrational lines observable through the 20th
harmonic. These lines display complex structure, including peaks (distinct from that of the Q-branch) that can be regarded as band heads.
Here we show how this spectral structure arises from vibration-rotation
interactions and vibrational anharmonicity.

I. INTRODUCTION

This paper concerns excitation of iodine vapor with argon ion laser
lines whose frequencies are higher than the convergence limit of the $3\Pi^+$
state (20 032 cm^{-1}). Such excitation within a dissociation continuum
produces a resonance Raman spectrum (RRS)[1,2]. Overtone progressions up
to the 20th harmonic can be detected. With narrow slit widths, we can
observe an apparent fine structure in all of the overtones[1]. Here we
summarize these results and show how the observed RR lines can be interpreted as S band heads and Q branches of vibration-rotation transitions
and corresponding vibrational hot bands. For the first time, band heads
can be found in a Raman spectrum.

II. THEORY FOR DIATOMIC MOLECULES

For a vibrational band the Raman displacement from the exciting
line for the Q ($\Delta J=0$), O ($\Delta J=-2$) and S branches ($\Delta J=+2$) are given by[3]

219

(neglecting centrifugal distortion terms):

$$(\Delta\nu)_Q = \Delta\nu_0 + (B_{v'}-B_{v''})J^2 + (B_{v'}-B_{v''})J \qquad (1)$$

$$(\Delta\nu)_O = \Delta\nu_0 + (B_{v'}-B_{v''})J^2 - (3B_{v'}+B_{v''})J + 2B_{v'} \qquad (2)$$

$$(\Delta\nu)_S = \Delta\nu_0 + (B_{v'}-B_{v''})J^2 + (5B_{v'}-B_{v''})J + 6B_{v'} \qquad (3)$$

where $B_{v'}$ and $B_{v''}$ refer to the rotational constants of the upper and lower vibrational state:

$$B_{v'} = B_e - \alpha_e(v' + 1/2) \qquad (4)$$

$$B_{v''} = B_e - \alpha_e(v'' + 1/2) \qquad (5)$$

The Raman displacement for the pure vibrational transitions $v'' \rightarrow v'$ are also given by[3] (neglecting higher order terms):

$$\Delta\nu_0 = (v'-v'')\omega_e - (v'^2-v''^2+v'-v'')\omega_e x_e$$
$$+ (v'^3-v''^3 + 3/2(v'^2-v''^2) + 3/4(v'-v''))\omega_e y_e \qquad (6)$$

Since α_e is positive, the quadratic ($B_{v'}-B_{v''} < 0$) and linear ($5B_{v'}-B_{v''} > 0$) coefficients in Eq. (3) have opposite signs, and therefore a maximum of $(\Delta\nu)_S$ is reached for a specific J value. Thus the S branches have band heads. The corresponding J_{max} values are calculated by setting $d(\Delta\nu)_S/dJ = 0$. This gives

$$J_{max} = \frac{B_{v''} - 5B_{v'}}{2(B_{v'}-B_{v''})} \qquad (7)$$

The Raman displacements for the S band heads for vibrational transitions $v'' \rightarrow v'$ are obtained by substitution of Eq. (7) in Eq. (3):

$$(\Delta\nu)_{S\ head} = \Delta\nu_0 + \frac{(B_{v''}-5B_{v'})^2}{4(B_{v''}-B_{v'})} + 6B_{v'} \qquad (8)$$

and by substituting Eqs. (4) and (5) in Eq. (8):

$$(\Delta\nu)_{S\ head} = \Delta\nu_0 + \frac{(4B_e-\alpha_e(5v'-v''+2))^2}{4\alpha_e(v'-v'')} + 6B_e-6\alpha_e(v'+1/2) \qquad (9)$$

III. EXPERIMENTAL RESULTS FOR IODINE

A survey spectrum of the overtone progression of the RR spectrum in iodine vapor is shown in Figure 1 for transitions for which Δv = 1 to 5 (Stokes) and Δv = -1 to -3 (anti-Stokes). Even though the slit width was 3 cm^{-1}, the rotational structure can already be recognized. When

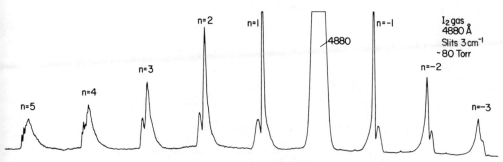

Figure 1. The resonance Raman spectrum of iodine vapor.

the slits are closed to 1.5 cm^{-1}, the fine structure of the bands can be clearly resolved (Figures 2 to 7). Figure 2 represents the region of the vibrational fundamental, whereas Figures 3 to 7 show the fine structure of the first, second, third, fifth, and eleventh overtones of the iodine vibration. For lower vibrational transitions, polarized Raman spectra were obtained. The parallel and perpendicular components of these Raman lines are shown in Figures 2 to 6. In the lower field of these figures the computer-calculated Fortrat diagrams of the Q, S and O branches are plotted for the first four vibrational transitions. Eqs. (1) to (6) and the spectroscopic constants given by Rank and Rao[4] have been used for that purpose. Vertical dashed lines in Figures 2 to 7 were drawn at the observed peak frequencies and serve as an immediate check on the assignments. The iodine fundamental consits of a relatively sharp unsymmetric line at 211.8 cm^{-1} and a weak broad line at about 227 cm^{-1} (see Figure 2). From the computer-calculated traces of the Δv=1 vibrational transitions we see that the line at 211.8 cm^{-1} is assigned to $Q_{v \to v+1}$ transitions. The Q branch traces are very steep at lower J values; therefore all the lines of a certain Q branch are very close to one another and give rise to an intense line. For the case of the fundamental of iodine, even the different $Q_{v \to v+1}$ transitions are too close to be resolved in our experiments.

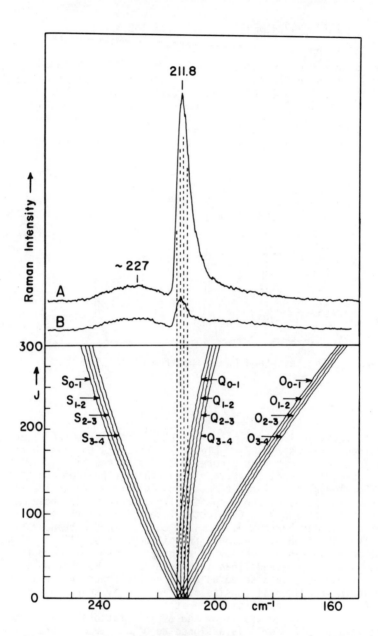

Figure 2.   Upper figure: Fundamental of the resonance
Raman spectrum of iodine vapor. Pressure
ca. 90 torr; 488 nm, 1 watt; slit width
1.3 cm^{-1}; A: $I_{\parallel}$, B: $I_{\perp}$
Lower figure: Fortrat diagram of the S,
Q and O branches.

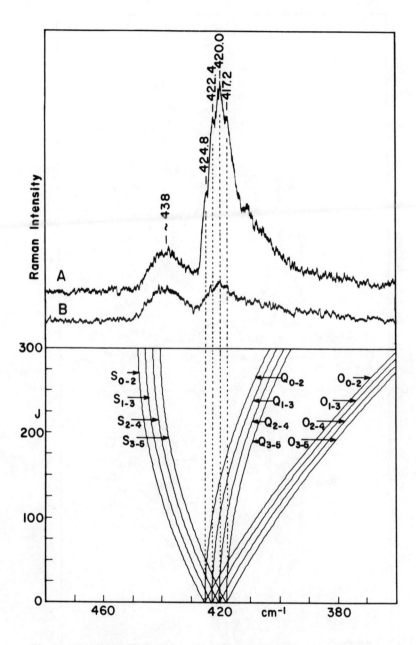

Figure 3.   Fine structure of the first overtone of iodine
vapor; caption as in Figure 2.

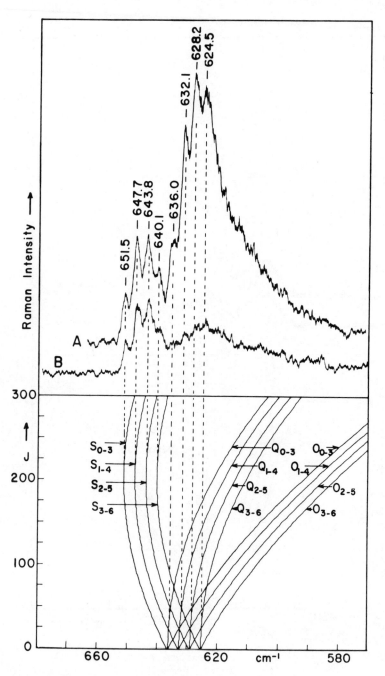

Figure 4.   Second overtone of iodine. Caption as in Figure 2 except for slit width 1.5 cm^{-1}.

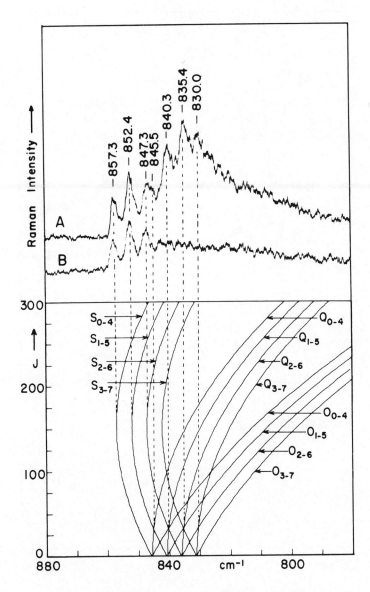

Figure 5.   Third overtone of iodine.   Caption as in
            Figure 2 except for slit width 1.5 cm⁻¹.

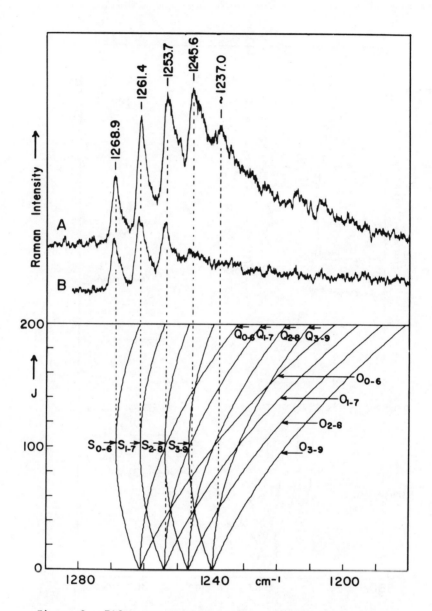

Figure 6.   Fifth overtone of iodine. Caption as in Figure
2 except for slit width 1.7 cm-1.

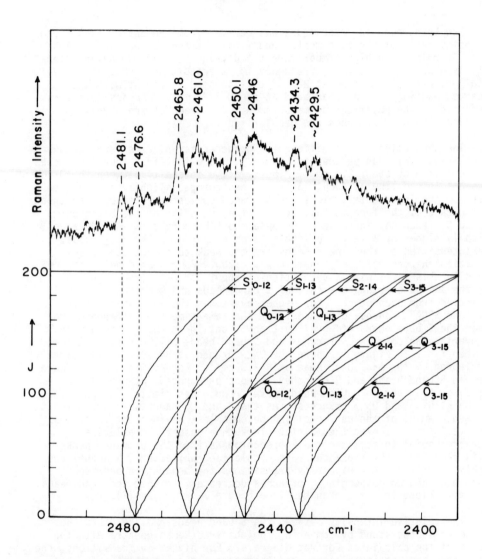

Figure 7. Eleventh overtone of iodine. Caption as in Figure 2 except for slit width 1.7 cm⁻¹ and no polarization spectra.

The S and O branches are not as steep as the Q branches for $\Delta v=1$. They are very much weaker since their lines are not superimposed. The weak broad line at about 227 cm^{-1} corresponds to the different S branches with $\Delta v=1$.

The spectroscopic temperature was derived from the anti-Stokes/Stokes intensity ratio of the fundamental and first overtone to be about 1150°K. For this relatively high temperature the J value for maximum Raman intensity is about $J_{max}=102$. The frequency for a transition $(v=0, J=102) \rightarrow (v=1, J=104)$ is calculated to be at 227.43 cm^{-1} in good agreement with the observed maximum of the broad band at about 227 cm^{-1}. The assignment is confirmed by the polarization measurements. Q branches are highly polarized, S and O branches are depolarized[5].

The separation of the different Q branches for $\Delta v=2$ is large enough to observe. This can be seen in Figure 3, where the first four Q branches are separated by about 2.4 cm^{-1}. The relative intensity between the Q and S branches (broad line at about 438 cm^{-1}) has changed slightly compared to that for the fundamental, since the S branch curves are steeper than those in the case of the fundamental, so that the intensity of the broad unresolved S branch has increased. Because the S band head is at a J value which is higher by a factor of about three compared to the J_{max} value, the intensities of the S band heads are too weak to be detected. This situation changes for $\Delta v=3$, where $J_{head} \approx 200$. Four S band heads can be found in this case, with a separation of about 3.8 cm^{-1} (see Figure 4). For the second overtone the sets of S band heads and Q branches are well separated from each other. For higher overtones the separation of the $Q_{0 \rightarrow v}$ and $S_{0 \rightarrow v}^{head}$ branches decreases. At the same time the differences between the S band heads of equal Δv transitions increase. Consequently, an overlapping between Q branches and S band heads will occur. This can be seen first in Figure 5, where the $Q_{0 \rightarrow 4}$ branch at 845.5 cm^{-1} is slightly overlapped by the $S_{2 \rightarrow 6}$ band head at 847.3 cm^{-1}. For $\Delta v=6$ the coincidence between S band heads and Q branches is more pronounced (Figure 6). Here the $S_{1 \rightarrow 7}$, $S_{2 \rightarrow 8}$ and $S_{3 \rightarrow 9}$ band heads have frequencies similar to those for the $Q_{0 \rightarrow 6}$, $Q_{1 \rightarrow 7}$ and $Q_{2 \rightarrow 8}$ branches. As the S and Q branches interpenetrate, the measurement of the peak position becomes less obvious until at $\Delta v=6$ the overlapping is almost complete. In such a case the depolarization data are useful to separate the S branch peaks from the Q branch peaks. It is clear that the S branches are depolarized while the Q branches are highly polarized. For higher Δv transitions the separation between S band heads and corresponding Q branches decreases, so that the band heads come very close to the Q branches (see Figure 7, where $\Delta v=12$).

The relative intensity between the S band heads and the Q branches can be well understood by comparing the observed Raman spectra with the traces of the calculated Fortrat diagrams. For higher Δv transitions, the Q branches are getting flatter, and therefore their intensities decrease slightly. On the other hand the S branches have their band heads at J values which are close to the J value for maximum intensity (at 1150°K, $J_{max}=102$). The greatest value of the ratio of the S band head intensity to that of the Q branches should be for $\Delta v=6$ and $\Delta v=7$, since there the

J value which corresponds to the band head is close to about 100. For higher vibrational transitions the intensity of the S band heads still decreases relative to the Q branches, because the S band heads occur at J values which are increasingly lower than J_{max}, so that few J transitions have frequencies within the spectral slit width.

The observation of many vibration-rotation transitions allow the determination of the spectroscopic constants ω_e, $\omega_e x_e$, $\omega_e y_e$, B_e and α_e with fairly high accuracy[1]. The constants derived from resonance Raman spectra are in good agreement with those obtained by fluorescence measurements[4].

IV. CONCLUSION

The observed fine structure in the resonance Raman spectra of iodine vapor is satisfactorily interpreted as series of S band heads and Q branches. Calculated Fortrat diagrams permit an easy assignment and some speculations about the relative intensities. The most striking features in the observed resonance Raman spectra for higher Δv transitions, however, are the S band heads.

ACKNOWLEDGMENT

The author is grateful to Dr. H. J. Bernstein (National Research Council of Canada, Ottawa) in whose laboratory the work was done.

REFERENCES

1.   W. Kiefer and H. J. Bernstein, J. Mol. Spectry. 43, 366 (1972).

2.   W. Kiefer and H. J. Bernstein, J. Chem. Phys. 57, 3017 (1972).

3.   G. Herzberg, Molecular Spectra and Molecular Structure I. Spectra of Diatomic Molecules, 2nd ed. (D. Van Nostrand Co., Inc., Princeton, 1950).

4.   D. H. Rank and B. S. Rao, J. Mol. Spectry. 13, 34 (1964).

5.   G. Placzek, Handbuch der Radiologie, Vol. 6, Part 2 (1934).

KIEFER DISCUSSION

BARRETT - How did you measure the temperature of 1150°K? Did you use 4880 Å irradiation?

KIEFER - Yes, we used the 4880 Å laser line. The spectroscopic temperature was determined by measuring the anti-Stokes/Stokes intensity ratio of the Raman bands. We also checked the temperature using the measured intensity distribution of the pure rotational spectrum of nitrogen which was added to the iodine.

BARRETT - Have you tried the 4965 Å line?

KIEFER - Yes, we also tried the 4965 Å line. This laser line also excites a resonance Raman spectrum. The 5017 Å line, however, gives rise to a resonance fluorescence spectrum on the Stokes side, but a resonance Raman spectrum on the anti-Stokes side, showing similar patterns with S-band heads and Q-branches. This can be easily explained due to the fact that the intermediate Raman level lies below the convergence limit for transitions which start at $v'' = 0$, but above the convergence limit for transitions originating from $v'' = 1$ (anti-Stokes).

SMITH - Do you expect to see any quenching of the Raman effect at extreme pressure?

KIEFER - There should be no quenching for the Raman effect, insofar as I know. The intensity should increase in proportion with pressure.

REMOTE RAMAN SCATTERING PROBES

by

S. H. Melfi

NASA-Langley Research Center
Hampton, Virginia*

ABSTRACT

One of the promising methods of remotely sensing molecular concentrations in the atmosphere is Raman LIDAR. LIDAR is an acronym for LIght Detectection And Ranging. The principle is similar to RADAR; a laser is used as the source of radiation and an optical telescope as the receiver. In this talk I will describe several field LIDAR systems and the results they provide. In particular, I will point out the advantages provided by Raman LIDAR observations for measurements of atmospheric transmissivity, particle loading, and concentrations of trace constituents.

* * * * * * * * * * * *

This talk concerns a field system for atmospheric probing that uses many of the processes discussed in previous talks. An artist's conception of this LIght Detection And Ranging (LIDAR) system is shown in Figure 1. The acronym LIDAR correctly implies a similarity to radar. The laser transmits a short pulse of light into the atmosphere. Particles and gas molecules scatter some of this light. A small fraction of the scattered light is collected by the telescope objective. We can analyze the spectral distribution of the collected light to get information about particle distributions, gas constituency and gas temperature, and the time dependence of this light provides range information.

Figure 2 illustrates the spectral distribution of a typical scattered light return from the atmosphere when the transmitted light is at 694.3 nm.

---

*Present address: EPA NERC-LV, P. O. Box 15027, Las Vegas, Nevada 89114

Figure 1.

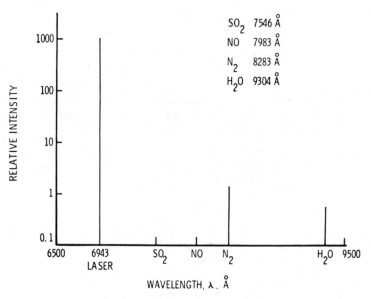

Figure 2.   Raman Return Wavelengths

The largest portion of the scattered light is unshifted in frequency.
This return is produced by Rayleigh scattering from molecules and scattering from particles. On the other hand, Raman scattering produces bands
of light shifted from the excitation frequency by amounts equal to the
vibration and rotation frequencies of the gas molecules. The figure
shows those bands produced by vibrational Raman scattering from nitrogen,
oxygen, water vapor, nitric oxide and sulphur dioxide. In this talk, I
want to concentrate on the importance of Raman scattering observations
in three types of LIDAR measurements: 1) measurements of atmospheric
transmissivity, 2) particle scattering, and 3) concentrations of trace
constituents.

First, let me describe the LIDAR system we have developed at the
NASA Langley Research Center. A photograph of this system is shown in
Figure 3. The most obvious component is the 48" Cassigranean telescope

Figure 3.

used to collect the scattered light. The small microwave disk is part of
an auxiliary radar system that is used for aircraft surveillance to insure
safe operation. Also visible are the two lasers used to generate the
transmitted light pulses. The ruby laser provides light at 694.3 nm with
which we use a doubler to get light at 347.2 nm. The Nd:glass laser
provides light at 1060 nm and, doubled, at 530 nm. Thus we can transmit
light at four wavelengths. This multiplicity of available wavelengths
is very useful in studies of atmospheric transmission and particle

scattering. The optical system is housed in the thirty foot van shown
in Figure 4. The second van in the photograph houses our data acquisition and computer analysis equipment.

Figure 4.

In operation, we fire the laser into the atmosphere, picking off a
small fraction of the transmitted light pulse for a photodiode monitor.
The monitor provides a measurement of the total pulse energy and a
"zero time" reference. The returning scattered light is collected and
collimated by the telescope. The collimated light is passed through one
or more interference filters, which select the spectral range of interest,
and then the light in this spectral range is detected by a photomultiplier.
Two modes of signal analysis are available. If the signal is strong it
can be sent through an amplifier to an analog to digital converter, and
then into a buffer memory and the computer. If the signal is weak, it
is analyzed by photon counting. Two scalars are used, with one counting
while the other reads out. This approach provides maximum time resolution
of 1 μsec. We can communicate with the computer by card reader, magnetic
tape and teletype, allowing a substantial amount of data analysis in real
time.

This system has just recently become operational. Before I discuss
the data we have obtained with it, I would like to show you some atmospheric
measurements we obtained with a much smaller system that has been operating
for several years. The smaller system uses a 16" telescope and the

original data were recorded on oscillograms with subsequent analysis by
hand. Figure 5 shows typical results for Raman scattering from nitrogen
and water vapor. In this case the transmitted light was at the doubled
ruby wavelength, 347.2 nm. The system was aimed at the zenith and data
was taken for altitudes up to 3 km. These results represent averages
over eight shots for the unshifted scattering and six shots for the
Raman scattering. The signals fall off rapidly with altitude because
of the $1/Z^2$ dependence which has been removed from the data by forming
the quantity $Z^2 V/E$, where V is the signal voltage and E is the laser
energy.

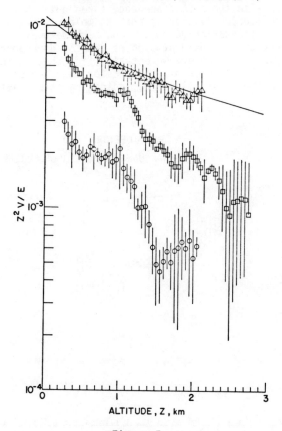

Figure 5.

One of the first things we wanted to check with regard to the Raman scattering measurements is whether the filters provide adequate rejection of scattered light in other spectral regions. In particular, the ratio of transmission at the Raman wavelength to transmission at the unshifted wavelength must be greater than 10^6 in some cases, especially if the air is dusty or cloudy. In order to obtain this strong rejection ratio we use two filters in parallel, carefully mis-aligned so as to reduce reflections back and forth between filters. In order to demonstrate that our rejection ratio was sufficient in these measurements, we fired the system when there was low cloud cover. Only in the unshifted scattering did we see the large return characteristic of cloud signals. In both the nitrogen and water vapor Raman signals we saw no increase at the cloud altitude and a substantial abrupt drop beyond. The drop results from attenuation of the incident and scattered beams by the cloud.

The result described above not only demonstrates that sufficient discrimination is available; it also illustrates the principle of atmospheric transmission measurements using Raman scattering. Now I would like to describe these measurements in some detail. Figure 6 shows the nitrogen Raman return with the $1/Z^2$ dependence taken out. Since the intensity of Raman scattering is proportional to molecule number density, if we do not have any attenuation going out and back we would expect the data to follow closely the standard atmosphere density dependence, which is also shown in Figure 6. However, we actually find that in this case, which is typical, the signal falls off much faster than the standard atmosphere density in the first kilometer, indicating strong attenuation

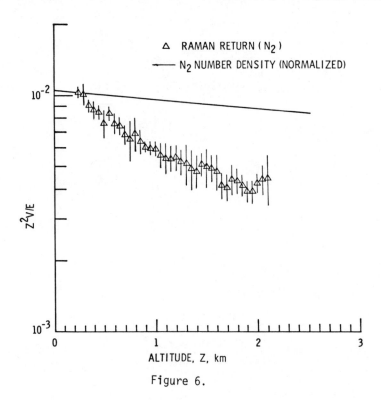

Figure 6.

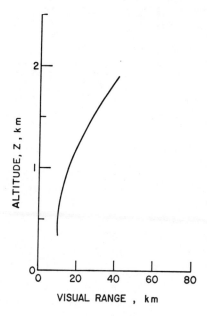

Figure 7.

in this region. At higher altitudes the signal follows the standard
atmosphere slope closely. Thus, from such measurements, we can get atmospheric attenuation as a function of altitude. In many cases attenuation
can be related to visual range, i.e., the range at which a dark object
can be seen against the horizon. In Figure 7 we show visual range as
a function of altitude, calculated from such data. In the illustrated
case, the visual range was quite small near the ground, on the order of
10 km. But above the mixing height the range increased to between 40
and 60 km.

Generally, atmospheric transmission cannot be measured using
Rayleigh scattering because scattering from particles appears at the
same wavelength, contributing an unknown background that is difficult
to separate and can vary rapidly. However, Raman scattering can be
separated from the particle scattering because of its spectral shift.
Thus, transmission measurements are provided by a comparison of the
density profile and corresponding Raman scattering from a major constituent such as N_2. Although it is typically three orders of magnitude
weaker than Rayleigh scattering, Raman scattering from N_2 has been observed to 40 km altitude by M. J. Garvey and G. S. Kent at the University
of the West Indies in Kingston, Jamaica. Thus it is by no means an
impossible measurement.

Observations of Raman scattering can also be used to facilitate
measurements of scattering by particles in the atmosphere. The vibrational Raman return from N_2 allows calculation of the corresponding
Rayleigh scattering contribution. Then the calculated Rayleigh contribution can be subtracted from the total unshifted return, leaving
the contribution from particles.

We are very interested in using the subtractive principle in making measurements in the stratosphere, in order to obtain a baseline against which to measure the effect of high flying aircraft. We first tried comparing the total unshifted LIDAR return from the stratosphere to a standard atmosphere molecular scattering profile. Unfortunately, independent measurements of molecular density from a balloon have shown that the deviation from a standard atmosphere in the stratosphere is often larger than the particle scattering. To illustrate this problem, Figure 8 shows the deviation of molecule number density from the standard

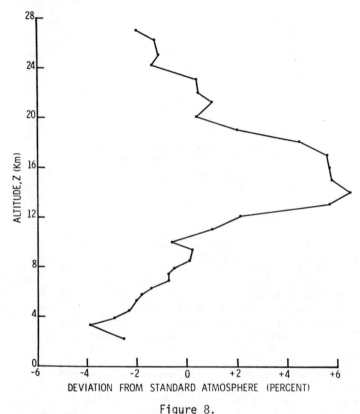

Figure 8.

atmosphere prediction on one night when we were taking both LIDAR and balloon data. Notice that in the region from 12 to 20 km there was a 2 to 6% increase. Without this result from balloon measurements, it would be tempting to interpret the consequent increase in scattering as a contribution from particle scattering. Observation of Raman scattering by N_2 can provide the same information as the balloon measurement; thus we expect that eventually a LIDAR system will provide an independent particle scattering measurement in the stratosphere.

For the present, the balloon flights provided other information that is valuable in demonstrating the utility of the LIDAR measurements. The balloons, which were launched by the University of Wyoming, carried particle counters that measured the particle density as a function of altitude. If the scattering properties and particle size distribution are constant over the observed altitude range, we would expect close agreement between counter and LIDAR particle measurements. Figure 9 shows a typical comparison. The LIDAR results were obtained with the 48" system described earlier. As you can see, the comparison is encouraging. If this agreement is confirmed in additional measurements, then eventually the LIDAR technique, using Raman scattering for density profiles, may prove to be a valuable tool in providing the numerous samples at different times and locations needed for good baseline information on stratospheric aerosols.

Now I would like to discuss remote Raman scattering measurements of trace constituents in the atmosphere. These measurements can be obtained by comparing Raman scattering from the trace constituent to Raman scattering from nitrogen. Quantitative results are calculated from this ratio by _in situ_ calibration against another measurement technique or by a

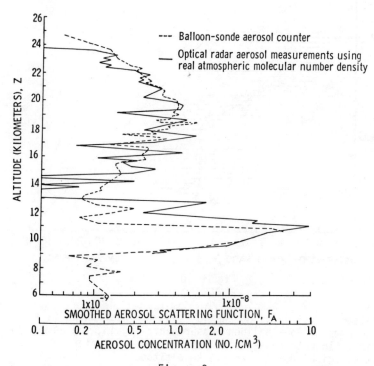

Figure 9.

system calibration and relative Raman cross sections, which have been
measured with good agreement by several groups. In the ratio measurement
the optical properties of the atmosphere cancel to the extent that they
remain constant over the small spectral interval between Raman lines.
Thus we do not have to make absolute scattering intensity or atmospheric
transmission measurements. For example, the ratio between the H_2O and
N_2 Raman signals from any altitude should be closely proportional to the
water vapor mixing ratio, which is a quantity of major interest.

In Figure 10 is shown a LIDAR measurement of water vapor in terms of
mixing ratio as a function of altitude. This result was obtained in

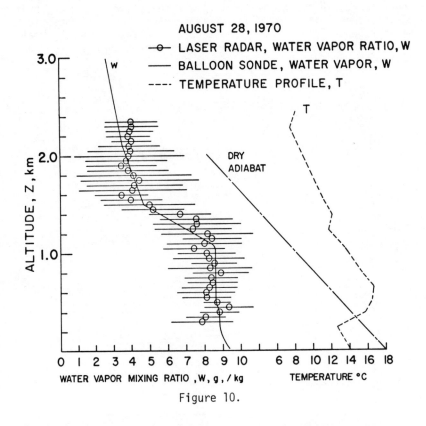

Figure 10.

Oregon's Willamette Valley using the 16" system. Also shown is a
corresponding simultaneous balloon measurement. Again, the agreement
is encouraging. In both cases there is a decrease in water vapor
content in the region from 1.1 to 1.6 km altitude. This behavior
corresponds very nicely to the prevailing temperature inversion, indicating that in fact there was a capping of mixing near the inversion.

Now I would like to describe another type of trace measurement, that
of sulphur dioxide in the exhaust plume of a power plant stack. To make

these measurements, we used another system with a 24" collector and a
ruby laser transmitter, shown in Figure 11. The scattered light was

Figure 11.

analyzed by an interference filter arrangement, which I will describe
shortly, and then detected by a photomultiplier. The photomultiplier
signal is pre-amplified and then fed sequentially to 15 counters to
provide range resolution. The counters are activated in time channels
that can be set as narrow as 10 nsec. We accumulate data by firing the
laser a large number of times, and then print out the total count from
each counter.

The problem of interference from particle scattering at the unshifted
wavelength is extremely severe in this application because of the fly ash
present in the plume. Consequently, we had to check very carefully to
make sure that our filters block the unshifted component. To perform
this check, we had two filters constructed, one of which used two separate
filter elements. Laboratory measurements showed that each element of
the composite filters provided a factor of 10^5 blocking at the laser
wavelength. The two elements were assembled with a 1° wedge angle between
them to reduce multiple reflections. The composite filter should provide
then nearly 10^{10} blocking. The other filter was composed of a single
element providing about 10^6 blocking. We made one measurement with the
composite filter and then very quickly made another with both filters.
The signal was reduced in the second case only by the transmission of
the second filter at the SO_2, ν_1 line. Thus we concluded that the blocking provided by the first composite filter is adequate.

Figure 12 shows a typical measurement obtained from a power plant
plume. The plume was located approximately 210 m from the LIDAR system,

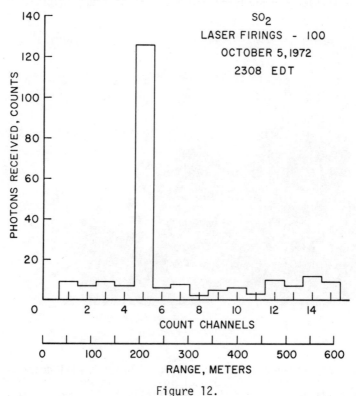

Figure 12.

corresponding to counter #5 of our data acquisition system and this is the channel in which we recorded a relatively large signal. The other channels, 1 through 4, and 6 through 15 provided a measure of average background. The results shown in this figure are derived from the accumulated count of 100 laser firings, in which the laser pulses varied in energy from 1 to 2 joules. Thus we are talking about between 100 and 200 joules of laser energy needed to get this type of signal. We also measured the nitrogen signal. Taking into account the relative SO_2 to N_2 cross section and system parameters measured in a separate calibration, we calculated that the observed SO_2 concentration was about 800 ppm. This result is in the range estimated from independent instrumentation mounted in the stack. Thus we have a fairly strong indication that we are actually seeing SO_2 in the plume. In a series of measurements, we observed a linear relationship between the apparent SO_2 signal and generated power as shown in Figure 13. This result is consistent with expectations provided that the efficiency of the plant and sulphur content of the coal remain constant over the observation time.

To summarize, I have discussed some ways in which observations of Raman scattering with a LIDAR system can be used to provide or facilitate remote measurements of atmospheric transmission, particle scattering and trace constituents. Although the Raman scattering is relatively weak,

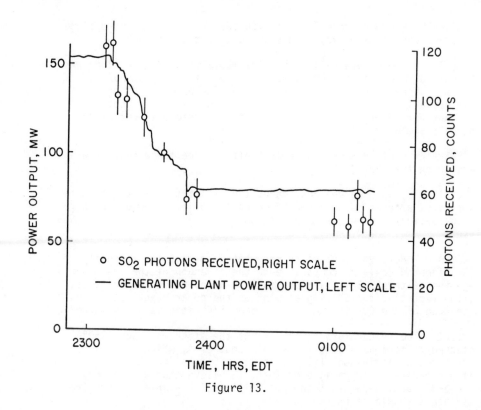

Figure 13.

it is by no means impossible to observe, and the information it provides is a valuable addition to that obtained from the stronger unshifted scattering.

MELFI DISCUSSION

KIEFER - Don't you think that the use of a small and high speed monochromator may be more efficient than using filters?

MELFI - It is very difficult, as I think was pointed out yesterday, to match a circular image from the telescope to a thin slit.

KIEFER - Even if you use image slicers?

MELFI - Yes, I still don't think you would get quite the transmission.

SCHILDKRAUT - There are six or seven other problems involved in using a monochrometer and two of the most important ones are, of course, the throughput and the depth of field of sample that can be imaged through the slits. When you figure it out, that depth is inconveniently shallow. And that means that regardless of what your gate time is, it is only a certain amount of the sample that you are seeing, limited by geometric optics. It doesn't matter how long you leave your photomultiplier tube on. That is something most people have neglected.

BERSHADER - I was wondering, did you observe the integrated plume in the SO_2 study, or were you focussing on a partial volume?

MELFI - We were integrating over the whole volume. The plume was only about 10 meters in diameter, so the total time that we had to observe the signal was on the order of about 100 nanoseconds or so. And, we actually couldn't accept much more than one or two photons during that period of time. We actually had to attenuate our signal coming back somewhat.

BERSHADER - Did you have an opportunity to see the effect of wind dispersion on the plume?

MELFI - No, we observed just about the lip of the stack, and this was our first attempt just to see if we could do it. It would be a very fine application later on, I think, to go down wind and see how the SO_2 diffuses.

LEONARD - Have you checked to see whether or not the intensity of the CO_2 sideband could be comparable to the strength of the SO_2 signal through the filters? This is a possibility at elevated temperatures. Also, the fact that the SO_2 signal went as the power plant output would also be true of the CO_2 if fuel of the same S/C ratio is being burned.

MELFI - The temperature at the top of the stack was about 250°F. Our stacked filters have a halfwidth of about 17 Å, which results from putting together individual filters of 20 - 25 Å halfwidth. Considering the line separation, which is over 70 Å, I believe that we are okay in this regard, but I agree that the possibility of CO_2 interference from the sidebands should be checked carefully.

STEPHENSON - Monitoring the CO_2 stack gas concentration would also help in generally characterizing the power plant operation. It does surprise me that the SO_2 concentration would go up as the power of the plant increases, when you simply expect to be pumping more air through. I would guess that the SO_2 concentration would stay roughly the same.

MELFI - Well, as a matter of fact, the Environmental Protection Agency made some stack gas measurements a few days before we made ours, and they also found a nearly linear increase of SO_2 with power. The relationship of SO_2 concentration with plant power can be truly complicated, though, and there are some older generating plants where the concentration would remain essentially constant.

LAPP - Why didn't you use the doubled ruby wavelength for the stack measurements?

MELFI - The answer is based upon our estimates of the tradeoffs in the system design. At the doubled ruby wavelength, the scattering cross section is larger by a factor of 16 and detector efficiency is also substantially better. The composite increase is much larger than the loss sustained in doubling. But it is more difficult to make adequate

interference filters for that spectral region, and fluorescence from
stack gases, particles and even from optical components is likely to be
a more serious problem there. We considered the various alternatives
carefully, and concluded that 694.3 nm was the wavelength of choice.
But filters are improving steadily, so at least regarding that aspect
of our choice the question is not finally settled.

PENNEY - Speaking of fluorescence, is it possible that part of the
signal at the SO_2 wavelength is fluorescence from stack gases or particles?

MELFI - Yes, that may be the case. Our tests so far do not rule out
fluorescence interference. During one of our experiments the precipitator
were shut down, and consequently the particle concentration in the plume
increased substantially. Our signal did go up slightly in this case.
One possible source of this increase is Raman scattering from sulphur
absorbed on the particles. By the way, most stack monitors filter out
the larger particles before measuring gaseous components, so they may be
indicating a lower total sulphur level than that actually exhausted.
This is an interesting possibility that must be examined more closely.

On the other hand, since our time resolution was about 250 nsec per
channel in these experiments, we could identify relatively long-lived
fluorescence because it would show up in channels 6 through 15. However,
the background in channel 6 was no higher than expected from preceeding
and subsequent channels. Nevertheless, the possibility of fluorescence
interference requires further examination, which might take the form of
two channel measurements, with one channel centered on the SO_2 line and
the other on an adjacent spectral region.

HENDRA - I believe I can contribute something with regard to the
possibility of fluorescence interference because at Southampton we have
been studying the Raman spectra of absorbed species on high surface
area materials. The types of absorbents we use are silica, alumina,
silica-alumina, zeolites, magnesia, titania, etc. i.e., oxides not
unlike those you are likely to encounter in a smoke plume from a coal
or oil burning furnace.

The major technological problem in this game is the occurrence of
high levels of fluorescence which tend to swamp out the Raman radiation.
The origin of this fluorescence is far from understood but its properties
are well documented. Typical characteristics are as follows:

1. The oxides are often non-fluorescent when cold, but heating to
    activate them for our work causes the fluorescence to develop.
    If we plot the level of fluorescence against the temperature
    to which the oxide has been preheated, we find that the fluores-
    cence increases up to a temperature of about 500°C, but then
    decreases sharply for preheating temperatures greater than
    900°C. It has also been reported by Sheppard and co-workers
    that heating in pure oxygen reduces fluorescence.

2. The fluorescence appears to the red of the exiting line as expected but also, to a lesser extent, to the blue.

3. The fluorescence decays rapidly with time when the sample is exposed to the laser beam but since it can often start out in excess of 1000 times the intensity of the Raman lines this effect is not adequate to remove it. The decay rate seems to be roughly proportional to laser power, and blue lasers are more effective than red in producing both fluorescence and its decay. The fluorescence intensity seems to rise, like Raman scattering, with the laser power.

It is widely held that the source of this nuisance is absorbed organic rubbish which is itself fluorescent.

In furnace-chimney combinations it is quite possible, in fact, that ideal situations present themselves to:

1. Generate a high surface-area impure oxide which could well be mildly fluorescent itself.

2. Heat it into the temperature regime where adsorption and "cracking" on the surface will be maximized.

3. Allow adsorption as the gas stream cools to < 200°C of any high boiling species (large unsaturated molecules in particular) which might be present in the gas stream.

If this is so, then one might expect strong fluorescence from the particles. In addition, if manganese or iron is present, there can be spiky fluorescence on top of any continuum from these impurities.

BLACK - Is it possible with these narrow filters to just tilt them far enough off the ν_1 Raman return of SO_2 to check the flourescence?

MELFI - You can tune them, and in fact Don Leonard used that technique to tune on and off the nitrogen and the oxygen lines in the very early work of Raman scattering in the atmosphere. But I am afraid of what that would do to the blocking with this small signal.

SALZMAN - Would you have enough return signal intensity to make a measurement on the anti-Stokes side so that the problem of filter fluorescence could be minimized?

MELFI - I doubt it.

BARRETT - Would you describe the characteristics of the ruby laser that you used for the SO_2 study?

MELFI - It was operated Q-switched, with 1 - 2 joule pulses of about 80 megawatt peak power and 25 - 30 nanosecond pulse duration.

SMITH - Unless I have missed a point here, and unless the environmental people are truly worried about how far the stack is from you, one
might be much better off not Q-switching, getting ten times the number
of photons out of a solid state laser on each pulse.

MELFI - That introduces an interesting trade-off. Its true that
you get more photons, but then its also necessary to leave the receiver
open a much longer time. If the background radiation is significant,
you may well lose by not Q-switching.

SCHILDKRAUT - We have looked at that trade-off in great detail,
and as a result we always try for a short pulse; 30 nsec is just what
you want.

GOULARD - In your Oregon measurements of atmospheric density profiles
[Applied Optics 11, 1605 (1972)] you indicated that there are two
potential sources of fluctuation in the results. One was the "weak
signal" statistics of Raman scattering; the other was due to the
possibility of density fluctuations in the atmosphere between pulses.
Did you evaluate the relative importance of these contributions?

MELFI - No. It is very difficult to estimate the magnitude of the
density fluctuations. As a matter of fact, it may be possible to use
the Raman technique to measure them. I think we can do that when we
take our next series of measurements with the large 48 inch system.

GOULARD - What was the time interval between pulses?

MELFI - For the original Oregon experiments it was 30 seconds. For
our 48 inch system it will be 1 pulse per second.

GOULARD - So, for the Oregon shots, a complete measurement required
about 5 minutes?

MELFI - Yes, typically.

RAPID AND ULTRA RAPID RAMAN SPECTROSCOPY

by

Michel Bridoux, André Chapput, Michel Delhaye,
Hervé Tourbez and Francis Wallart

Université des Sciences et Techniques
de Lille (France)

Presented by Professor M. Delhaye

ABSTRACT

An electromechanically scanned monochromator is used for rapid
recording of Raman spectra from gases and liquids. This system has
demonstrated its usefulness in applications to Raman spectroscopy of
time-evolving samples. The limitations of the electromechanically
scanned system are discussed and a new electro-optic system is described
that utilizes an image intensifier and television camera tube to simultaneously record spectral information in a large number of channels.
Results with the electro-optic systems are presented.

I. INTRODUCTION

In work performed at our laboratory, we have found that high
quality Raman spectra of gases and transient samples can be obtained
in times on the order of one second or less. These measurements are
very useful in the study of transient and unstable samples. We have
investigated the use of two different techniques: rapid scanning and
ultra-rapid multichannel recording. This communication describes our
implementation of these techniques and some of the results we have
obtained. More details are found in Refs. 1-3.

II. RAPID SCANNING EXPERIMENTS

For rapid scanning we have used a double monochromator to which
has been added a high speed scanning device. Useful spectra can only

be obtained under the condition that noise is kept low and that the
signal is of reasonable strength.

The spectrometer consists of a rotating helical cam, which moves the
cosecant bar drive in an oscillating motion. A repetitive scanning that
is linear in time is thus obtained. The spectral range of this scan can
be adjusted from 50 cm^{-1} to 1000 cm^{-1}, and the range can be positioned
anywhere in the visible spectrum.

The signal is observed on an oscilloscope, whose horizontal sweep is
generated by a rotating potentiometer linked to the scanning cam. This
feature permits a wave number calibration to be obtained during each
scan. The rapid spectra are usually stored by means of a magnetic tape
recorder, or by means of a rapid strip chart recorder. Various examples
of spectra obtained under these conditions using an argon laser source
are described below.

A. Chlorine Data

The isotopic structure of the vibrational line of Cl_2 is well resolved
in a series of rapid spectra recorded in 0.2 second from a sample in a
solvent, or in gas phase. The quality of these spectra is sufficient to
allow detection of hot bands and isotopic species during chemical reactions.

B. Bromine Reactions

The rapid scanning apparatus permits a kinetic study of chemical
reactions in the liquid phase. For example, we consider the observation
of bromine, bromine complexes and the effects of catalysis during organic
reactions resulting from a bromuration on carboxylic acids such as acetic
acid. The "active" form is not the diatomic Br_2 molecule, but "associated"
forms such as $n(Br_2)X^-$ in which the bromine molecules are distorted and
complexed. We have studied the Raman spectra recorded by rapid scanning
in the low frequency range for different complexes of bromine in water or
in acetonitrile, which are obtained by mixing bromine with various anions
(fluoride F^-, cloride Cl^-, ...). The acceleration of reactions with
increasing temperature, and the influence of catalysis by CH_3-COBr have
been observed. These experiments have been compared with other techniques
such as conductimetry, and it has been shown that the scattering data
provide additional information. The activation energy was calculated from
Raman results, yielding about 5 kilocalories.

These rapid spectra were obtained from very small samples. The
technique can also be used to study phase transformations in solids,
especially in microcrystals.

C. Raman Study of Atmospheric Gases

Using a cw argon laser source and the rapid scanning apparatus, we observed both the pure rotational, and the rotational-vibrational bands of the Raman spectra of atmospheric gases. Instead of using a small gas cell placed in a multipass optical system, we have designed a spectrometer for studying the "free" atmosphere. A schematic diagram of this device is shown in Figure 1. The laser beam is directed along

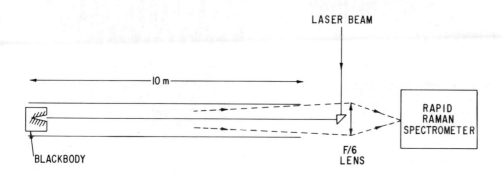

Figure 1

the axis of a black plastic tube, whose length is approximately 10 meters. A black cone absorbs the laser light at the end of the tube. The back-scattered light is collected by an f/6 lens, which focuses it onto the slit of a rapid scanning spectrometer. Figure 2 shows several examples of rotational Raman spectra of air and pure gases at normal pressure obtained using this arrangement. Figure 3 shows the vibrational bands of oxygen in air, of CO_2 at natural concentration in air (~300 ppm), and of the C-H band of the vapor of an organic compound added to air.

The increase in intensity of the CO_2 bands in air due to human breathing is easily detected and can be measured with fairly good precision.

We are now extending this technique to remote detection of pollutants in the free atmosphere. For example, the products of various combustion reactions are detectable.

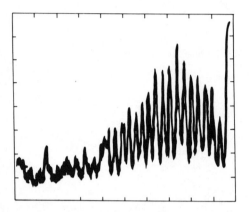

Fig. 2a.   Pure rotational Raman spectrum
of CO_2.   Exciting line = 514.5
nm; power = 1 w; scan repetition
time = 10 sec.; slit width =
1 cm^{-1}.

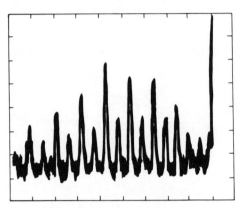

Fig. 2b.   Pure rotational Raman spectrum
of N_2.   Exciting line = 514.5
nm; power = 1 w; scan repetition
time = 10 sec.; slit width =
1 cm^{-1}.

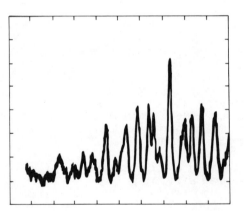

Fig. 2c.   Pure rotational spectrum of air.
Exciting line = 514.5 nm; power
= 1 w; scan repetition time =
10 sec.; slit width = 1 cm^{-1}.

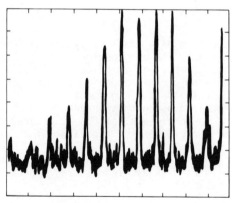

Fig. 2d.   Pure rotational spectrum of O_2.
Exciting line = 514.5 nm;
power = 1 w; scan repetition
time = 10 sec.; slit width =
1 cm^{-1}.

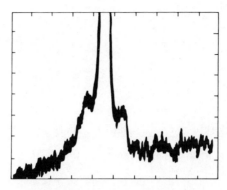

Figure 3a.   Vibrational spectrum of $CH_3 CO CH_3$ (5% in air).
Exciting line = 514.5 nm; power = 1 w; scan
repetition time = 10 sec.; slit width = 20 cm^{-1}.

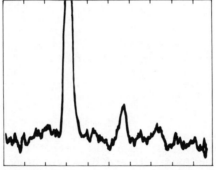

Figure 3b.   Vibrational spectrum of CO_2 (300 ppm in air).
Exciting line = 514.5 nm; power = 1 w; scan
repetition time = 15 sec.; slit width = 20 cm^{-1}.

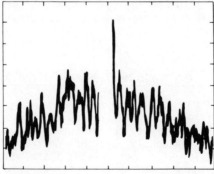

Figure 3c.   Rotational-vibrational spectrum of O_2 (in air).
Exciting line = 514.5 nm; power = 1 w; scan
repetition time = 15 sec.; slit width = 4 cm^{-1}.

D. Static Samples

The use of data storage and averaging by means of a small digital
computer permits a dramatic improvement of sensitivity for use with non
time-evolving samples. For instance, C_6H_6 at low concentrations ($\sim 10^{-5}$)
in various solvents and in vapor phase are well resolved by averaging
400 rapid spectra, in a total time of 400 seconds.

E. Limitations of the Rapid Scanning Technique

The main problem with rapid scanning, which is now conventionally
used for the study of many chemistry problems, is that great care is
required to obtain acceptable signal-to-noise ratios. The noise is
relatively high because each spectral channel is observed for only a
very short time. Thus, if the scan width is S and the required spectral
resolution is $\Delta\lambda$, then $S/\Delta\lambda = n$ channels must be observed, one at a time
in the total measurement period T. Typically, $S/\Delta\lambda$ is on the order of
100 to 1000 and T is near 1 second. Consequently only a few milliseconds
are available for each spectral channel. During this time $\Delta t = T/n$,
only a small number of photons will be detected per channel in typical
cases. If the average number of detections is N, then the percent statistical fluctuation among a series of measurements will be approximately
$(1/\sqrt{N})$ x 100%. This fluctuation appears as noise on the spectrum, and
obviously, as the observation time per channel is reduced, N also gets
smaller and the noise is increased.

In general, the noise associated with monochromator scanning tech--
niques is greater than optimum because only one spectral channel is viewed
at a time, while light in other channels, though carrying useful information, is blocked by the exit slit and lost. In order to reduce measurement times, we have investigated a second technique that avoids this loss
of information by viewing each of the spectral channels for the full
measurement time. This "multichannel" technique is described below. The
most interesting case concerns the recording of a rapid spectral event
during a short time $\leq \Delta t$ where the scanning technique is evidently inoper-
ant, while the "multichannel" technique is successful.

III. ULTRA RAPID MULTICHANNEL TECHNIQUE

The multichannel technique is based on the use of image intensifier
phototubes and a high sensitivity television camera. This combination
is able to record, simultaneously, all the spectral elements of the
spectrum. This basic layout of our present instrument is shown in
Figure 4. The radiation focused on the entrance slit is dispersed in a
first optical Czerny-Turner configuration by a plane grating. In the
focal plane of the second mirror, we place a wide adjustable slit which
limits the spectral band pass to the desired spectral range. This slit
acts as an entrance slit for a second Czerny-Turner grating spectrograph.

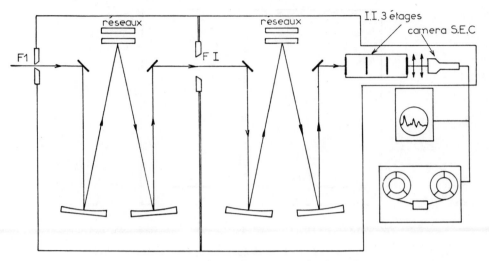

Figure 4. Multi-channel Electro-optic Spectrometer

All the optical elements of the spectrum are then focused on to the photocathode of a three stage image intensifier. The intensified spectral images which appear on the phosphor screen of this tube are transferred by means of high aperture lenses to the photocathode of a SEC television tube. The optical image is transformed into an electronic image which is integrated in a multiplying and storage KCl target. This target is able to integrate the signal from nanoseconds to minutes. The integrated information is read by means of a beam of slow electrons as in conventional television camera tubes. The video signal picked up on the signal plate can be used to produce an image on a television monitor, or it can be analyzed on a CRT or transferred to a magnetic or electronic multichannel memory for further treatment of information.

The use of the integration capability of the SEC camera allows analysis of Raman spectra excited by a single laser pulse. All the information is simultaneously recorded during the pulse duration (typically from microseconds to milliseconds). Using this system with a "giant pulse" ruby or YAG, we can obtain a complete Raman spectrum in ten or twenty nanoseconds. Thus, for example, in Figure 5, the complete Raman spectrum of $GeCl_4$ is recorded with good signal-to-noise ratio. The incident light energy required to obtain this spectrum was only 160μJ! In Figure 6, we show vibrational Raman scattering from three simple gases, each obtained in less than a microsecond using 8 mJ of incident light.

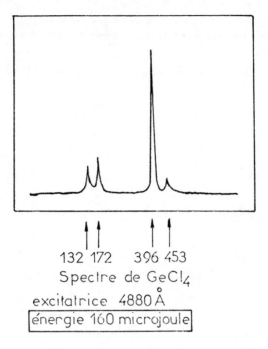

132 172   396 453

Spectre de GeCl$_4$

excitatrice 4880 Å

énergie 160 microjoule

Figure 5.

This early work on multichannel Raman data acquisition clearly indicates its utility for the recording of the spectra of atmospheric gases using a single laser excitation pulse. It is now being extended through use of both pulsed ruby and nitrogen lasers.

REFERENCES

1.   M. Bridoux, Rev. d'Optique 8, 389 (1967).

2.   Michel Delhaye, Appl. Opt. 7, 2195 (1968).

3.   M. Bridoux, A. Chapput, M. Crunelle, and M. Delhaye, in Advances in Raman Spectroscopy, Vol. 1, ed. by J. P. Mathieu (Heyden and Son, Ltd., London, 1973) Chapt. 7.

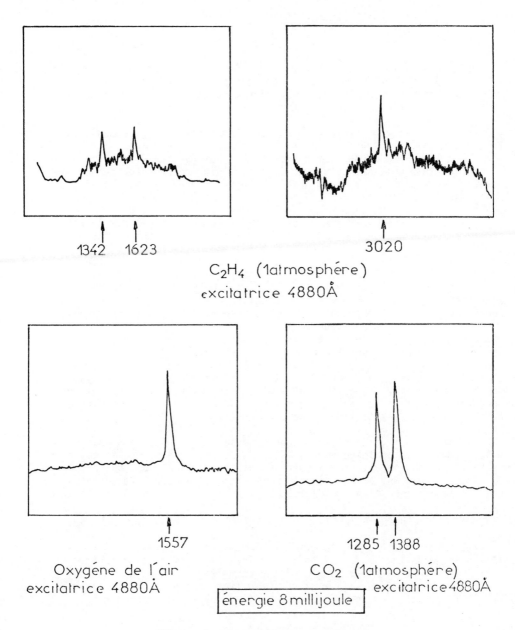

Figure 6.   Raman Spectra of Gases

DELHAYE DISCUSSION

KIEFER - I think everybody was very impressed about the speed with which a Raman spectrum can be obtained with your method - especially if one considers how many hours a conventional Raman spectroscopist spends to obtain just one good Raman spectrum.

BARRETT - If you take a good photomultiplier and look sequentially by the conventional scanning method at each spectral element for a certain amount of time, one second for example, then how does the signal-to-noise in the spectrum obtained by conventional scanning compare to the signal-to-noise in a similar spectrum obtained in one second by the use of your multichannel technique?

DELHAYE - The comparison is very difficult because we have to consider, on one hand, the noise for the total area of the photocathode of the photomultiplier, and on the other hand, the noise from a very small area of the photocathode of the image intensifier which is the area of one spectral line. From that point of view, the image intensifier is far better than the photomultiplier, because the probability of thermal emission of electrons is lower. In practice, we may select a very good photomultiplier and it is not so easy to select a good image phototube. Theoretically, the gain is due to the number of spectral elements which are recorded simultaneously, and that gives a decisive advantage to the image intensifier in the case of spectra excited by short pulses.

BARRETT - Yes, but your signal-to-noise ratio is governed by your overall sensitivity. I guess what I am really asking is: how does the spectral sensitivity of the image intensifier with your system compare to the spectral sensitivity of the photomultiplier?

DELHAYE - Well, the photocathode is exactly the same in a photomultiplier and in an image intensifier phototube. The signal-to-noise ratio is due not only to the properties of the photocathode, but also to further treatment of the signal. It seems to be difficult to integrate during a long time the signals arising from an intensifier phototube.

BARRETT - The gain is higher in a photomultiplier.

DELHAYE - You are right, but you cannot easily compare the photomultiplier with the image intensifier. The image intensifier is theoretically able to achieve single photon counting. With a good quality three stage image intensifier, a photoelectron produces a light signal on the fluorescent screen which is detectable by an SIT camera tube.

ELECTRONIC SIGNAL PROCESSING FOR
RAMAN SCATTERING MEASUREMENTS

by

E. Robert Schildkraut

Block Engineering, Inc.
19 Blackstone St.
Cambridge, Mass. 02139

ABSTRACT

General signal processing techniques for laboratory Raman spectroscopy
and Remote Raman spectroscopy using cw and pulsed lasers are discussed.
Chopped and pulsed source/time gated detection techniques are outlined.
Detectors such as photomultiplier and image tubes, SEC Vidicon sensors,
image dissectors, and multiplex spectrometers are discussed and some of
their relative merits evaluated. Computer aided data reduction and techniques for fluorescence suppression in the signal and in the display are
covered.

A brief comparison of signal processing systems for Remote Raman,
Micro-Raman (spectra of single particles one micron in diameter) and
other high background instrumental configurations are examined. Suitable
references are detailed for further study of each of these topics.

I. INTRODUCTION

A general treatment of electronics for Raman signal processing would
encompass all of the common and most of the exotic signal processing
techniques in use today. In order to avoid such a lengthy and inappropriate
paper, I will cover here a few selected case histories and refer liberally
by reference to other examples or expansions on particular techniques.

In the following discussion we should try to keep in mind four
relatively independent factors related to signals and noise. These are
common to most signal processing problems but are particularly relevant
in the case of Raman spectroscopy and are not always understood. These
factors are:

1) average signal level,

2) average noise level,

3) dynamic range of the signal, and

4) expected dynamic range of the noise.

In designing the optimal signal processing system for each application, all of these factors must be evaluated in order to select the
right combination of detector, preamplifier, amplifier, and post processing chain for best signal display and interpretation. Raman spectroscopy
of exotic samples which, for example fluoresce, is an extremely challenging
problem and most of the gains made in these "techniques" over the past
several years have been due primarily to judicious and careful application
of known signal processing techniques. New tools such as mode locked
lasers, as they become available, enable a wider variety of these
techniques to be applied, but a basic understanding is necessary in
order to implement them correctly.

II. PHOTON COUNTING AND SYNCHRONOUS DETECTION

In laboratory Raman spectrometers, two primary modes of signal
detection are used, depending upon the detector and the particular
nature of the signal or signal source. Photon counting, which is a
relatively noise-free process, is used in the presence of weak signals
where relatively low background levels are present. In the case of flame
spectroscopy, the Raman spectroscopy of highly fluorescent biological
samples, or systems in which self emission is a problem, synchronous
detection using chopped sources is the more commonly employed technique.

In photomultiplier tubes, which are the most commonly used detectors
for Raman spectroscopy in the 2500 Å to 9000 Å region, almost all of the
noise present in the signal is either statistical noise caused by the
discrete nature of the arriving photon flux or is "first dynode noise".
The photon noise is proportional to the square root of the number of
photons arriving per second. The only true noise added by the photomultiplier tube is due primarily to the statistical fluctuation in gain
between the cathode and the first dynode caused by a variety of photomultiplier fabrication and electric field artifacts (geometry of the
dynode, arrival position of the photoelectron, and non-unity quantum
efficiency of the cathode). Photon counting techniques usually enable
operation of the detection system over an extremely wide dynamic range
covering the region from the arrival of a few photons per second up to
several hundred thousand photons per second before bunching makes the
individual photon events coalesce. There are tubes with inherently
linear operating regions of twelve orders of magnitude in anode
current! Not all of this dynamic range is usable by subsequent
amplification or display devices however, and optimization of the match
between sensor and display has been traditionally given less than needed
consideration.

Commercial systems[1] have become available for convenient use of
pulse counting advantages. Direct reading pulse rate meters combine
the S/N advantages of pulse counting and the convenience of an analog
(meter) or digital display. Careful selection of the best operating
point for each PMT is necessary in order to optimize the performance
of these systems.[2] Above certain critical voltages, within the tubes'
operating limits, afterpulsing, ion feedback, and anode feedback increase
disproportionately and reduce the signal/noise at the output.

Although not all of the multiplier noise comes from the first
dynode, the contributions of the latter (higher current) dynodes become
successively less important as we move up the chain toward output end
(anode).[3] As the gain of the first stage increases, the contribution
of the latter stages to the noise is decreased still further.[4] A
stable high gain first dynode therefore can reduce the overall "noise
added" by the PMT to the inevitable photon noise. The new high gain
GaP tubes enable just such low noise operation and have an increased
pulse height discrimination ability that allows differentiation between
1, 2, 3, 4, and more simultaneous photoelectron events. This is very
important in applications such as Remote Raman where pulsed lasers
are used and photon counting in the normal sense cannot be accomplished
during the short (tens of nanoseconds) pulse.

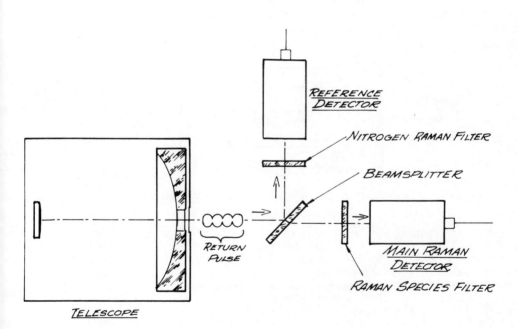

Figure 1.   Pulsed laser Raman scatter return from a remote volume.

III.   REMOTE RAMAN SYSTEMS

Photons arriving back at a gated detector, Figure 1, come in a
"bunch", so densely packed that the interval between photoelectrons
is too small to resolve. If however the PMT linearly sums the multiple
photon events, the total charge delivered to the anode will be directly
proportional to the energy in the return pulse and we can in effect
"count" the photons or photoelectrons ex-post-facto.

One convenient method for doing this, which is compatible with
subsequent digital processing, is through the use of a gated digitizer.
Such a process, common to nuclear instrumentation, is implemented as
shown in Figure 2. The anode pulse is gated into the digitizer by

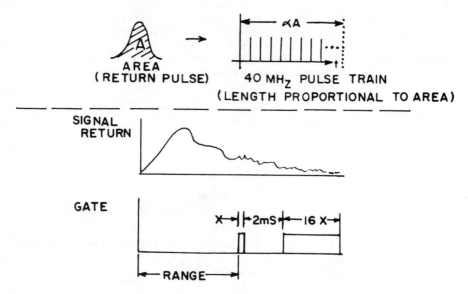

Figure 2.   Schematic of gated digitization process.

appropriate control electronics which select the right time for sampling
the signal return. The digitizer then starts to emit a series of pulses
in a synchronous train, the length of which is proportional to the area
under the gated input pulse. The number is accumulated in a storage
register as a count corresponding to the pulse area or total photon
flux. This digital data is more easily processed, since background
subtraction and ratio operation are normally part of a well designed
remote Raman system. The dynamic range of typical gated digitizers
is 200:1, which is only a fraction of the linear operating region of
even the poorest PMT, so either special digitizers should be built or

one must center the operating region of the PMT/digitizer combination
such that the expected signal returns make use of as much of the linear
range as possible. This can be accomplished by changing the gain of
the PMT (varying its dynode string voltage) or by incorporating, between
the PMT and digitizer, a variable gain preamplifier. Both schemes
are less than optimum, but building wide range gated digitizers is not
a trivial task either.

One method for changing the gain of a PMT which suffers from fewer
disadvantages than most others is to vary the potential on only one (1)
of the dynodes - preferably an intermediate one. PMT's are carefully
designed to operate optimally with a well defined, stable electrostatic
field distribution in the multiplier section. The electron optics are
usually severely degraded (especially at the photocathode and near the
anode) if the first and/or last dynode potential fields are distorted.
Intermediate dynode collection and multiplication are less severely
affected, and since any dynode can effectively "cut - off" the gain of
the tube, choosing an intermediate dynode for gain control is appropriate.[5]
It is also much easier and faster to change the potential on one dynode
than to switch the whole string even if the individual dynode time
constants are short (they are usually not).[6] Gain can easily be varied
over a factor of 100 with a potential swing of only tens of volts.
In some cases, this can be done fast enough such that a "constant
amplitude" display (A scope) of a returning remote Raman signal could
be shown on a CRT screen. One would start the detector tube at low
gain, trigger the CRT sweep and increase the gain with time, precisely
the amount needed to counteract the $1/(\text{range})^2$ dependence of the returning
signal. The signal/noise would deteriorate toward the right hand side of
the display but the amplitude everywhere would be a meaningful quantitative representation of the level of Raman return (photons). The voltage
waveform necessary to achieve such dynamic gain control is not particularly
exotic and can be achieved with a minimum of passive components and one
fast electronic switch. An electronic system approximating this
performance in discrete range increments is shown schematically in
Figure 3. Here, it is anticipated, quantitative data at only fifteen (15)
contiguous, discrete ranges are needed and the Raman signal over each
range increment is averaged and displayed in a histogram-like format.

After a remote Raman signal is recorded in a storage register,
the detector is turned on again a short time later (2 msec.) to record
the background at the same wavelength. This is then subtracted from the
first (Raman plus background) signal to yield a Raman photon flux value.
Since remote systems should be useful in all weather, day and night,
the dynamic range of background signals and their noise extends over
nine (9) orders of magnitude typically. To reduce the noise in the
background signal somewhat more than this simple approach allows,
it has been suggested by Hirschfeld[7] that the background be integrated
for about 16 time periods, the result divided by 16, and the quotient
subtracted from the Raman plus background for each pulse. Extending
the background integration beyond 16 periods is not productive in view
of the noise remaining in the Raman signal itself.

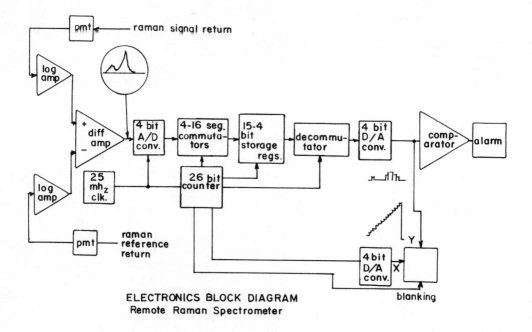

ELECTRONICS BLOCK DIAGRAM
Remote Raman Spectrometer

Figure 3.   Electronic system for A-scope display of Raman return.
This system corrects for $1/(range)^2$ dependence of backscatter.

After background subtraction, a ratio is taken between the "pure
Raman" signal from the remote sample and the Raman signal from nitrogen
in the same volume, at the same time. This ratio reduces the effects
of laser pulse height variation, atmospheric transmission, scintillation
and sample inhomogeneity. The result of all this careful processing
can be seen in Figures 4 and 5 which represent before and after such
processing respectively. In this case, a nitrogen Raman spectrum was
referenced (ratio taken) to the laser Rayleigh backscatter rather than
to itself.

In this application, the dynamic range of background (10^9), dynamic
range of its noise (10^4:1), dynamic range of the expected signal, and
noise in the signal must all be evaluated so that the system operating
point can be optimally set.

A more exotic extension of such electronics takes three (3) samples
in series when measuring stack effluents (Figure 6) and can thus
partially compensate for the plume's opacity and particulate content.

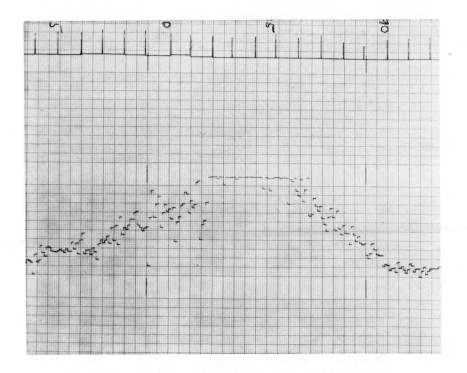

Figure 4.   Raw data remote Raman signal for N_2 (count rate vs. wavelength).

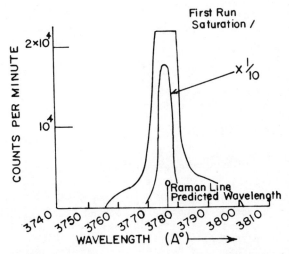

Figure 5.   Processed remote Raman signal for N_2 (count rate vs. wavelength) divided by laser Rayleigh backscatter signal.

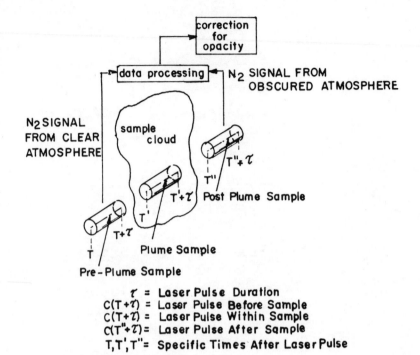

τ = Laser Pulse Duration
$C(T+\tau)$ = Laser Pulse Before Sample
$C(T+\tau)$ = Laser Pulse Within Sample
$C(T''+\tau)$ = Laser Pulse After Sample
T, T', T'' = Specific Times After Laser Pulse

Figure 6.   Schematic of remote Raman system for sampling three
volumes (pre-plume, plume, and post-plume) used to
compensate for plume opacity and particulate content.

IV.   MULTIPLEX SPECTROMETERS

One technique for increasing the utility and power of Raman detection
systems is the use of multiplex techniques. Looking at all of the wavelengths simultaneously enables a significant reduction in time to obtain
a complete spectrum.[8] In some cases (transient events, Micro-Raman spectroscopy, fugitive or delicate samples) this gain in efficiency is the
only way one can obtain a spectrum at all.

Fourier Transform Spectroscopy, which immediately comes to mind, is not a clearcut present solution in conjunction with Raman spectroscopy for a variety of reasons which will not be covered here. (N. B. Editors: This topic is covered by Dr. Schildkraut in another contribution to this Proceedings.) Fabry-Perot interferometry is being used successfully by Barrett et. al.[9] to study the pure rotational Raman spectra of various compounds, but this is not a universal tool. (N. B. Editors: See the presentation by J. J. Barrett in these Proceedings.) Since we generally rely on a multiple dispersive spectral instrument of some kind for our wavelength sorting, the presence of an input slit or aperture, and its inherent throughput limitation, will be considered as given for this discussion.

The subject of image tubes in general and their application to multiplex spectroscopy[10] in particular is a complex and difficult one for a variety of reasons, not the least of which are manufacturers' erroneous or more insidiously irrelevant claims. Image tubes are manufactured primarily for the image industry. The eight shades of gray obtainable on a good home TV set and controlled studio lighting do not warrant the use of a wide dynamic range camera tube. The resolution required of some military systems does maintain the high resolving power technology in many of these tubes, but only two or three shades of gray (or green) are needed to visualize a sniper among trees, etc. Hence the orthicons, vidicons, isocons, silicon vidicons and their intensifier cousins are, in general, suitable only for restricted spectroscopic applications and then should be chosen with great care. It would be presumptuous to attempt here a matrix selection table which could be used to make a choice of tube for each application and the references should be consulted if use of an image tube is contemplated. It will help to point out a few typical problems associated with their use and some schemes for partially circumventing their limitations.

To treat the problem in a coherent manner it is best to describe a case history in which a multiplex Raman spectrometer, operating in the visible region was an absolute necessity. The discussion will treat part of a Micro-Raman spectrometer which should be capable of obtaining the Raman spectrum of a single particle 0.1 μm or larger. The very small Raman photon flux emanating from such a particle when it is illuminated by as intense a beam as possible (below the photodecomposition level), and the limited time available in which to obtain a spectrum (again due to decomposition), dictate the need for a multiplex Raman spectrometer to make the device feasible. Although an immense array of PMT's could be used, some image tubes can closely approximate this performance if handled correctly and represent a considerable cost saving, not to mention the ability to capture an essentially continuous spectrum without the discrete separations inherent with the multiple PMT approach. In order to format the spectrum for best use of the image tube <u>area</u>, a crossed dispersion echelle spectrograph is used (Figure 7) which produces a very high dispersion spectral array, slices it into discrete portions, and stacks these orders one over the other in a raster-like format, uniquely suited to the image tube readout. The format of the spectral display is shown in Figure 8.

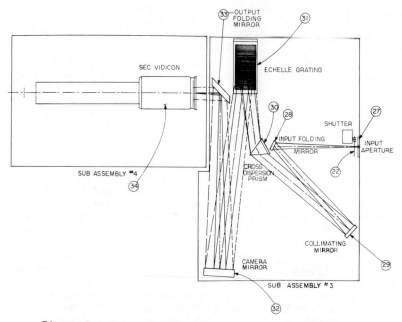

Figure 7.   Crossed dispersion echelle spectrograph.

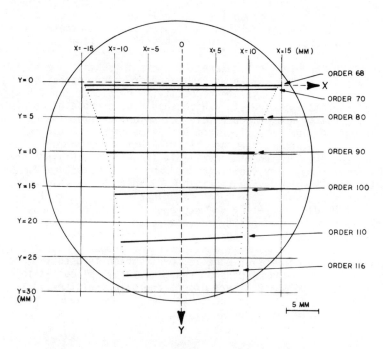

Figure 8.   Spectral map of the image at the exit focal plane of the
crossed dispersion echelle spectrograph shown in Figure 7.

The image tube could be a vidicon, orthicon, isocon etc. but for optimum dynamic range, sensitivity, low noise and good storage characteristics, a Secondary Electron Conduction (SEC) vidicon was chosen. This tube has a standard photocathode followed by a special target electrode in which electron multiplication takes place; each image element is amplified and the resultant charge pattern is stored on a target for subsequent readout in the Micro-Raman system by a digitally controlled electron gun. The target has also the properties of an almost perfect integrator of charge over a reasonable dynamic range and hence is the equivalent of an array of PMT's and capacitors - 500 wide by 500 high. It has low equivalent dark current and can store the charge for many minutes without leakage error. Each spectral element occupies an area only 30 x 30 μm on the faceplate and hence the photocathode dark current (proportional to photocathode area) per channel is much less than would be obtained with a standard PMT for each wavelength.

A short digression on image tube dynamic range is in order at this point. Manufacturers' claims for image orthicons and vidicons call out dynamic ranges on the order of 50-300. In most cases these are obtainable only under conditions which are never obtainable in real applications. The SEC vidicon claims a dynamic range of 300:1 and one can actually obtain 100-200:1 in practice. Orthicons usually obtain 100:1 and in TV applications rarely need more than 30:1. The orthicon however has a negative-white characteristic, that is, when no light is incident on the photocathode, the return (signal) beam current at the electron gun is a maximum. The current decreases as light increases on the photocathode. Hence, for zero spectral signal, one has maximum amplifier noise due to electron current shot noise (equivalent to photon noise statistically). The image isocon avoids at least this difficulty by maintaining a positive white output, where, for zero light input the anode current is minimum. The isocon has other problems related to dynamic range and noise injection due to the complex electron gun structure needed to strip the sheath electrons (signal) from the main return beam electrons (carrier) at the business end of the tube. All image tubes with storage targets suffer to some extent from a phenomenon known as lag or image sticking. This means that a complete discharge of the target is not achieved with one pass of the electron beam and hence photometric accuracy requires several scans. Other target materials and geometries have their advantages and drawbacks also and will not be discussed here; the references should help.

Since in most Raman spectra, the dynamic range displayed and encountered is several orders of magnitude, the detector must be at least capable of something on this order or better. The addition of intensifier sections on the front end of any image tube, although possibly increasing sensitivity (not always), almost invariably compresses the dynamic range available from the combined device. Intensifiers should be used with great care since they rarely help in spectroscopic situations where dynamic range is critical.

Micro-Raman spectroscopy[*] is just such a case, and a special process is used to extend the effective dynamic range of the tube without burning holes in its target. Since the target is a near perfect integrator, we allow weak spectral lines to integrate on it for longer periods of time before interrogating and reading them out with the electron beam. Strong lines (as determined by a rapid initial scan) are read more frequently. Computer control and digital, random access electron beam deflection enables such selective interrogation of the spectrum. The computer remembers how long it has been since each spectral position was last "read" and computes the intensity based on this factor as well as the charge measured. In this way, a storage target with an inherent dynamic range of only several hundred can be made to yield a true dynamic range of tens of thousands. The low lag and nearly perfect integration characteristics of the SEC Vidicon enable such a technique to be implemented. This scheme is the subject of U.S. Patent #3,728,576 among others.

In certain cases integration of charge at only certain locations (spectral points) on the faceplate of an image tube can yield a S/N improvement greater than $\sqrt{t}$ where t is the integration time of the photon flux.[11] All non-ideal photosensors and even phosphors (photographic film, photoemitters, photoconductors, etc.) have a typical transfer characteristic shown in Figure 9. There is always a

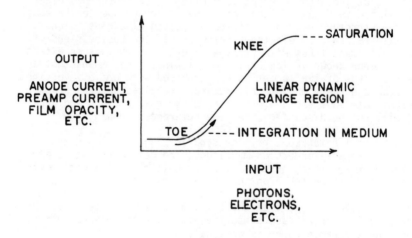

Figure 9.   Real device transfer characteristic.

*Work sponsored in part by Air Force Technical Applications Center, Patrick AFB, Contract #F33657-71-C-0679.

threshold, linear region of usable dynamic range, and a knee or saturation level above which output tapers off. The threshold is of key
interest in most cases and is partially a result of chemical and base
fog (photographic film), dark current (PMT's) or preamp and target
noise (most image tubes). In situations where we coherently add
signals hidden in the system noise, we can improve the S/N ratio by
$\sqrt{N}$ where N is the number of signal periods added or the number of times
a repetitive signal is coherently summed. The random noise accumulates
as $\sqrt{N}$ while the coherent signal adds linearly as N. The signals
to be added, however are always among or within the "toe" region of
the characteristic curve.

Prefogging photographic emulsions to move the input threshold
into the linear region is a well known technique and could in principle
be applied to a similar integration medium such as the SEC target.

The level of prefog would have to be very uniform over the tube
face and carefully controlled; it would also have to be known accurately
in order that the "prefog uncertainty noise" not do more harm than good.
Since prefog also tends to reduce the already limited usable dynamic
range available at the target, it is a viable technique only in certain
restricted cases where wide dynamic range is not needed or where
expected signal levels are known.

Integration within the medium (film, photo target etc.) rather than
at the output of the chain allows a better gain in S/N. Suppose we
have a nearly perfect integrator (surprisingly well approximated by a
SEC image tube target) of low leakage.[12] As our photon flux impinges
on the tube, the charge accumulates until we round the toe of the
curve and enter the linear region. For a given time period, this
yields a non linear incremental improvement in S/N, better than $\sqrt{t}$,
the usual integration time gain. This is true when the dark current
from the tubes' photocathode is not the limiting noise source (as with
the SEC tube where the preamplifier noise is predominant). The remaining
problem is how to read out this carefully accumulated charge, and
amplify it in the most noise free manner. Given state-of-the-art
preamplifiers with reasonable (1 MHz) bandwidths, one should measure
each image element in the minimum time such bandwidth allows. The
electron beam in these tubes is used to recharge or neutralize the image
charge built up on the target. It is this nonuniform recharge current
that yields the signal. Since the charge can be replenished using a
moderate beam current for a moderate duration or a high current for
short time, we choose the latter. In this way, the preamp is "on" for
minimum time and it contributes its constant noise current to the output
signal for a shorter period; the preamp noise contribution is thus minimized.

A SEC vidicon with dynode multiplier on the return electron beam
would be a more nearly ideal sensor since the dynodes could amplify
the output current level to the point at which the preamp noise was a
negligible factor. Such tubes, while they have been built, are not
generally available now and the future availability is not clear due to
limited demand in the market (which doesn't need this level of S/N).

The combination of random access to the faceplate, reading out strong spectral lines frequently (allowing weak ones to integrate on the target) and interrogating the target charge by short duration high current pulses, yields an optimum S/N situation for this type of visible multiplex television spectrometer. Intensities on the order of only tens of photons per second can successfully be detected in each spectral element in this way.[13]

V.   <u>FLUORESCENCE SUPPRESSION</u>

Fluorescence suppression schemes have proliferated recently and a word or two about these is relevant. Several attempts have been made to differentiate between fluorescence and Raman photons by using polarization, spatial, temporal, or statistical[14] differences in the signals; none have been completely successful. [N. B. Editors: The differentiation between fluorescence and Raman scattering is discussed elsewhere in this Proceedings, and, in particular, in the presentations of P. J. Hendra and C. M. Penney.] One technique which has potential (primarily because it has not been perfected yet) is the use of picosecond laser pulses and picosecond gated detectors. Since normal, non-resonance Raman signals are the result of essentially instantaneous (10^{-12} second) processes, they can be differentiated

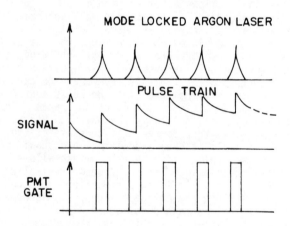

TYPICAL SPECS (LASER)   TYPICAL PMT GATES

PULSE INTERVAL: 7 nSEC   2 nSEC (10% POINT)
PULSE WIDTH: 180 Pico SEC
PULSE ENERGY TO: 10-100:1
    CW ENERGY

Figure 10.   Schematic of time gating for photomultiplier tubes in fluorescence suppression system.

from even short lived fluorescence (10^{-9} seconds) if adequate detector
gating can be implemented as shown in Figure 10. The gating of PMT's
or imaging tubes much below the one nanosecond region is difficult, if
not impossible at present, so a static optical gate of some type is
needed to take advantage of the very short laser pulses which emanate
from cw and pulsed mode locked lasers. Theoretical bandwidth limitations on the solid state lasers which produce the 6-10 picosecond
pulses make them relatively unsuitable for Raman excitation. Their
spectral bandwidth is rarely less than 200 Å. Continuous wave, mode
locked, argon lasers however do exhibit adequate spectral purity
and have pulse widths on the order of 100-300 p sec. This may still
enable adequate differentiation between Raman and fluorescent signals
from most of the more troublesome fluorescent biological samples.
We are presently working on techniques to enable ultra fast time gated
detection to be evaluated.

In Micro-Raman, we expect an unusual effect to become evident,
that of fluorescence saturation.[15] In the presence of the very high
incident flux (10^{16} photons $sec^{-1} \mu^{-2}$), we expect the cooled sample to
exhibit a fluorescent level essentially dependent upon the probability
of individual molecular de-excitation and emission. For the smallest
samples (0.1 μm^3) the photon flux is approximately 10^8 photons-
$molecule^{-1}-sec^{-1}$. It is likely then that any molecule in the ground
state will not stay there for very long. Hence we will have a
saturation of the particles' fluorescence. The Raman signal does not
saturate until much more light is incident on the sample[16] and hence
the Raman-to-fluorescence ratio in Micro-Raman is likely to be favorable.

The varieties and combinations of detection systems are almost end-
less. With great care, and attention to every detail of the processing
system choice as well as its construction, significant improvements can
be made over present quality displayed data. In some cases, the
care can be rewarded by the opportunity to make a new experiment
possible.

REFERENCES

1. SSR Model 1120 Amplifier-Discriminator and 1105 Data Converter
    Manual, Solid State Radiations, Los Angeles, Calif. 1969, pp
    2-1 through 2-10.

2. Ibid.

3. Simon Larach, "Photoelectronic Materials and Devices," (D. Van
    Nostrand Company, Inc., New Jersey, 1965).

4. RCA Photomultiplier Manual PT-61, RCA Electronic Components,
    Harrison, New Jersey 07029 (1970).

5. John Harding, Block Engineering, Inc., personal communication.

6. Robert J. Leser and Jack A. Salzman, "Light-Detection Electronics
    For a Raman Lidar," NASA TN D-6879 (1972).

7. Tomas Hirschfeld, et al, "Remote Spectroscopic Analysis of ppm-
    Level Air Pollutants by Raman Spectroscopy," Appl. Phys.
    Lett. 22, 38 (1973).

8. Michel Bridoux and Michel Delhaye, "Spectrométrie Raman-Laser
    Ultra-Rapide," Nouv. Rev. d'Optique Appliquée 1, 23 (1970).

9. W. H. Smith and J. J. Barrett, "The Use of The Transmission Windows
    of the Fabry-Perot Interferometer in the Detection of Raman
    Scattered Radiation for Atmospheric Gases," AIAA Paper, 71-1078.

10. Tomas Hirschfeld, "Raman Spectroscopy of Samples in the One Cubic
    Micron Range," Int. Conf. Raman Spectry., Ottawa, (August, 1969).

11. Tomas Hirschfeld, "Raman Microprobe: Vibrational Spectroscopy
    in the Femtogram Range," Opt. Soc. Am. Spring Meeting,
    Denver (March, 1973).

12. G. W. Goetze, et al, "SEC Camera Tubes," Reprinted from the
    publication, Advances in Electronics and Electron Physics,
    Volume 22, Photo-Electronic Image Devices, Third Symposium,
    (Academic Press, London). Available from Westinghouse Inc.,
    Elmira, New York.

13. Op. Cit., Hirschfeld, Ref. 11.

14. John Cooney, "Can LIDAR Detect CAT?" Part 2, Applied Optics
    11, 2374 (1972).

15. Tomas Hirschfeld, "Elimination of Fluorescence in Raman Spectroscopy,"
    Ann. Mtg., Opt. Soc. Am. (1973).

16. J.-P. E. Taran, ONERA, personal communication. (See also the
    presentation of J.-P. E. Taran in this Proceedings.)

SUPPLEMENTARY REFERENCES

Multiplex Spectroscopy:

A. J. F. James and R. S. Sternberg, "The Design of Optical Spectrometers,"
    (Chapman and Hall Ltd., London, 1969).

B. Robert John Bell, "Introductory Fourier Transform Spectroscopy,"
    (Academic Press, New York, 1972).

C. Manuel Cardona, "Modulation Spectroscopy," (Academic Press, New
    York, 1969).

Noise Analysis:

D. R. H. Brown and R. Q. Twiss, "Interferometry of The Intensity Fluctuations in Light," Proc. Royal Society, London 242A, 300 (1957); 243A, 291 (1958).

E. William H. Louisell, "Radiation and Noise in Quantum Electronics," (McGraw-Hill, New York, 1964).

F. C. D. Motchenbacher and F. C. Fitchen, "Low-Noise Electronic Design," (John Wiley & Sons, New York, 1973).

G. RCA Electro Optics Handbook, SCN 102-67, RCA Defense Electronic Products, Aerospace Systems Division, Burlington, Mass. (1968).

H. Athanasios Papoulis, "The Fourier Integral and Its Applications," (McGraw-Hill, New York, 1962).

Detectors-Imaging and Non-Imaging:

I. B. Kazan and M. Knoll, "Electronic Image Storage," (Academic Press, New York, 1968).

J. Daniel R. Frankl, "Electrical Properties of Semiconductor Surfaces," (Pergamon Press, New York, 1967).

K. R. G. Neuhauser and L. D. Miller, "Beam Landing Errors and Signal Output Uniformity of Vidicons," Jour. of SMPTE 67, (March 1968).

L. F. L. Skaggs, et al, "A Broadband Image Pick-Up Tube with High Near-Infrared Sensitivity," 1969 Electro-Optical Systems Design Conference.

M. Walter G. Jung, "Camera System Design Considerations Involved in Application of The Silicon Target Vidicon," Available from MTI Division, KMS Industries, Inc., Cockeysville, Md. 21030.

N. T. Kohler, et al, "The Silicon Diode Array Camera Tube for Near Infrared TV Applications," 1969 Electro-Optical Systems Design Conference.

O. A. H. Sommer, "Photoemissive Materials," (John Wiley & Sons, Inc., New York, 1968).

Examples of Spectroscopic Signal Processing Tricks:

P. H. D. Pruett, "Photon Counting System for Rapidly Scanning Low-Level Optical Spectra," Applied Optics 11, 2529 (1972).

Q.   Erwin G. Loewen, "Diffraction Grating Handbook," Bausch &
      Lomb, Inc., (1970).

R.   Potts, Speed Suppression Programmed Scanning, Dow Chemical Co.,
      ASTM publication.

SCHILDKRAUT DISCUSSION

DAIBER - I would like to ask a question about the real device transfer characteristics curve. (See Figure 9). What causes the toe in the
curve?

SCHILDKRAUT - This is the typical response one obtains from the
potassium chloride (KCl) target in the SEC vicicon. The toe is usually
caused by noise of various origins (preamp noise, dark current, etc.).

STEPHENSON - While we are on this, somebody has suggested that the
equivalent of prefogging can be accomplished. How does this work out?

SCHILDKRAUT - You can "prefog" the target, and you pay the price,
which is slightly compressed dynamic range. Also, prefog, in such a
sensitive detector, must be quite carefully controlled such that the
noise in the prefog level is small. This is non-trivial when you consider how few photons you need in each resolution element and calculate
the shot noise from point to point on the tube at this flux level.

PENNEY - How is the entrance slit in Figure 7 oriented?

SCHILDKRAUT - The "slit" if you were to examine it under a micro-
scope is a tiny square or rectangle aligned with the major dimension
vertical-across the spectral orders. The small dimension is along the
order, in the high resolution direction.

PENNEY - I see. The reason you can disperse in two directions in
this case is because you are using only a small fraction of the usual
slit height.

SCHILDKRAUT - That's right. There is no reason to use the full
height of the slit since the particle and its image are nearly circular.
For remote Raman too, a small circle would be more appropriate but is not
quite as efficiently stacked close together as the square. With crossed
dispersion one could, in some cases, use a higher slit and use fewer
spectral orders; a more dispersive prism would be needed in that case.
You must avoid the vertical overlap of orders. The prism dispersion
separates the spectral pattern vertically and the grating further dis-
perses each section laterally - that is what the raster-like pattern in
Figure 8 illustrates.

BERSHADER - Just out of curiosity, may I ask whether those two bright yellow lines belong to mercury?

SCHILDKRAUT - I don't know. Probably not, since they appear in the wrong place. My familiarity with photography tells me the color on the transparency need have nothing to do with the real color of the line for highly intense spots. This is what happens when you push color film hard enough. The film comes up with whatever color it wants to! Those could be any lines; as a matter of fact, they could be red since they are in the red orders. You've seen color photos of ruby laser beams on axis and the blooming yellow or blue-white pattern that obtains.

BARRETT - Do you see the hyperfine structure of the 5461 Å line of mercury?

SCHILDKRAUT - No, but we can see the fine structure of the laser and its spurious plasma lines. We therefore have a pre-monochromator or purifier in the Micro-Raman system which operates on an expanded spatially and spectrally filtered beam before it ever gets focused on the sample.

BARRETT - Just that one line?

SCHILDKRAUT - We have an argon-krypton mixed gas laser because we feel that probably the shorter wavelengths will destroy our sample and we want to be able to go to a progressively redder and redder wavelength. We thought originally that the problem in this instrument would be that we would create the world's best laser emission probe (i.e., would vaporize the sample, and we would get emission spectra). The customer we are building this for can't loose these particles, and must use non-destructive testing. He probably has only a few particles in existence. In any case, the limiting factor on lifetime for the particle in the laser beam turns out not to be thermal degradation (we are cooling the sample to liquid nitrogen temperatures), since we can maintain the heat rise to reasonable temperatures near 50 degrees C. based on absorbtion of typical samples. Two-photon chemical degradation is predicted to be the major problem. A 0.1 micron particle may contain only 10^8 molecules. For a 50 mw_o beam we have 1.2×10^{17} photons/second incident on the sample (at 4880 A). This is almost 10^9 photons per molecule per second. The probability of two photon events is high indeed. We have not seen this experimentally yet.

SMITH - Would you tell us how many lines of the scan of the SEC you are going to use for each dispersed segment of the spectrum?

SCHILDKRAUT - We will use as many of the SEC resolution elements as possible. We are trying to match it exactly. The tube is digitally scanned and hence can have up to 1024 discrete TV lines in principle. We will use one electron beam scan-line per order and anticipate using about sixty (60) orders initially since that is all we require to do a complete Raman spectrum. You will notice on the photograph of the actual pattern similar to Figure 8 that the orders are not uniformly spaced or completely straight. This is one reason for using digital scan; the scan line can be made to exactly follow the orders. The system is fully capable of scanning 500-600 orders if correctly designed. Blooming of a bright spot into adjacent vertical orders on the tube target must also be considered however.

SESSION IV
TECHNOLOGICAL APPLICATIONS
Session Chairman: D. Bershader

INTRODUCTORY REMARKS FOR SESSION ON TECHNOLOGICAL APPLICATIONS OF RAMAN SCATTERING, WITH EMPHASIS ON FLUID MECHANICS AND COMBUSTION

by

D. Bershader

Department of Aeronautics and Astronautics
Stanford University

I have the challenging task of trying to shepherd an attempt to assess some of the studies discussed here and the related progress in the light of technological applications, particularly for fluid dynamics and combustion research and testing. That subject area is used here in the broad sense and is meant to include related areas of pollution, propulsion, aeronomy studies, and possibly other areas of engineering interest to all of us here.

I have purposely not planned any rigid structuring for the session this afternoon because I feel there should be enough freedom for us to review subject matter in any way we wish to, and to develop some healthy controversial discussions as we go along. In this connection, we can, of course, amplify those aspects of the discussion which appear most fruitful. Also, it is my understanding that there are some persons here who have done very interesting work on these problems and who have not yet had a chance to communicate their results to the group. I would hope that these persons will contribute to our discussion as the opportunity arises.

From one point of view I may be a reasonable choice to chair this discussion in that I have no ax to grind, not having worked in Raman spectroscopy. My own special interest was to learn a little bit about the technique and what has been going on in the current research, and in fact, I feel I have learned a lot at this meeting. I will, however, take the chairman's perogative to make a few remarks which I hope will set the stage for the discussion to follow.

I believe that Raman spectroscopy can play an increasingly important role in the near future in the study of both fundamental mechanisms and engineering behavior in the field of pollution, fluid mechanics, combustion phenomena, interactions between laser beams and fluids, aerodynamic flow configurations, and aeronomy and atmospheric sciences. A problem common to these areas may be mentioned at the outset, namely, that of mixing processes. Here, one wants to identify ratios of species as well

as physical conditions such as temperature. One wishes also to look at
transport behavior of various physical quantities which would in turn
relate to the level of turbulence and reaction rates. Ultimately it
would be desirable to deduce both micro and macro mechanisms for turbulent diffusion, combustion reactions, and other processes, possibly
including wave propagation.

     Mixing processes and turbulent behavior continue to require atten-
tion as do current aerodynamics and aeronautical studies. Research
groups at places like the NASA Ames Research Center have developed a
renewed interest in turbulent boundary layer profiles in selected flow
fields. Vortices shed from big airplanes are now an important problem
again because their persistence has been shown to constitute a hazard
for smaller airplanes flying through the wake region even several minutes
later. Additional attention is being given to turbulent flows and engine
exhausts because of the problem of noise generated by aircraft. Apart
from the noise problem, there is a continuing interest in the familiar
problems of behavior of exhausts plumes from rocket nozzles at various
altitudes, both with respect to mixing and spreading behavior, and with
respect to content of chemical species and reaction rates. The environmental framework for these aerodynamic and propulsion phenomena requires
parallel studies in aeronomy and related atmospheric sciences, where we
want to identify species participating in thermo-convective types of
flows associated with pollution and meteorological phenomena.

     These days there is also considerable interest in the detailed nature
of flows associated with gas dynamic and chemical lasers and possibly
other exotic types of lasers as well. Here again, mixing processes
occurring in or near the optical cavity evidently play an important role
in the efficiency of energy conversion. It would appear that Raman
spectroscopy would be a powerful tool to give quantitative description to
the mixing and to measure new designs which would modify undesirable mixing features. In this overview of the problem areas, I have only mentioned some of the more obvious ones; I am sure that everyone here could
add to this list.

     To implement successful applications one must clarify the problem
definition by suitable scaling and modeling. In addition, one must, of
course, continually assess the capabilities of the experimental methods,
and ask whether it is feasible or practical to make improvements in
sensitivity or modifications in time response, etc., as may be required
for particular experiments. The question of modeling is a basic one
because if we do know the scaling factors and similarity laws describing
a particular problem, it becomes possible to conduct experiments under
controlled laboratory conditions which may be able to isolate individual
elements of the phenomenon under study. In turn, one is then able to
design a more intelligent full scale experiment, which would, among other
things, save a good deal of expense. In the case of turbulence and turbulent mixing, there is indeed a whole history of modeling and scaling,
both of the phenomenonological type as well as the statistical type. I
suspect that earlier progress, however, may not be adequate for the
applications of current interest, but there are persons here who are
more highly qualified than I am to make such judgements.

Whenever an exotic method is first applied in a subject area, the old and familiar questions have to be raised, the answers to which may sometimes prove somewhat embarrassing. For example, in a typical experiment, what is the size of the effective scattering volume, and in what way is the scattered signal averaged over the scattering volume? What is the intrinsic precision of the measurement technique, i.e., what is the fractional partial derivative of the measured quantity with respect to a change of the independent variable? Are we getting responses to more than one variable and can these responses be separated? One should ask hard-nosed questions such as whether laser scattering is better than a small resistance wire probe, or sodium line reversal, etc. for purposes of temperature measurement. It is desirable to show quantitatively, if it is indeed true, that in certain configurations of flows of particular species mixtures, Raman scattering, using available or developable light sources, gives better combined time and space resolution than other methods. There are also related practical questions: What is the ease of operation and transportation? What about the cost of the equipment and what about the lifetime?

This completes my set of introductory remarks which were intended, at the expense of some intrinsic coherence, to stimulate some lively discussion, and also intended to bring out additional information from people here who have their own experience to contribute. To add some further momentum to this session, we do have a few people with short sets of prepared remarks, but primarily what I hope this session will turn out to be is a good brainstorming interaction.

LASER RAMAN SCATTERING - A TECHNIQUE
FOR ARC-TUNNEL FLOW CALIBRATION

by

A. A. Boiarski[*]
Naval Ordnance Laboratory

Fred L. Daum
Aerospace Research Laboratories

ABSTRACT

Laser Raman Scattering (LRS) has been investigated as a possible
approach toward providing a nonperturbing flow calibration method for
arc-heated wind-tunnel facilities. Such a calibration is needed to
evaluate the chemical and thermodynamic properties of the high-temperature
air flow which cannot be accurately determined using standard probe
techniques. Theoretical calculations showed that the number densities
and vibrational temperatures of nitrogen, oxygen, and nitric oxide
could be obtained using the Raman spectroscopic method. Exploratory
experimental LRS results obtained in the Air Force Flight Dynamics
Laboratory 50-Megawatt Re-entry Nose Tip (RENT) Facility showed that
it is indeed feasible to apply LRS to arc-tunnel flow diagnostics.

I. INTRODUCTION

The diagnostics of high-enthalpy arc-heated wind-tunnel flows pose
many experimental difficulties, and therefore, an accurate calibration
of the operating conditions in these facilities is often lacking.

One problem arises from the fact that probes, when inserted
directly into the flow, disturb the fluid sufficiently to render the
results inconclusive. Hence, it is desirable to use some sort of
remote sensing method, foremost of which are the spectroscopic, interferometric, and microwave techniques. It is generally accepted that

[*]Dr. Boiarski was a visiting Research Scientist at ARL under Contract
F33615-71-C-1463 at the time the experiments were performed.

spectroscopic techniques are best suited for diagnostic work since the others can only ascertain gross properties of the gas, while spectroscopic analysis includes the possibility of examining individual constituents of a gas mixture.

Although promising, most spectroscopic techniques have various drawbacks which limit their use to certain molecular species and/or density regimes. Furthermore, some are path rather than point dependent. However, it can be shown that, utilizing the laser Raman scattering spectroscopic technique, point measurements of species concentrations can be determined for individual molecular constituents over a wide density range in a general gas mixture of aerophysical interest. Also, the static temperature of the gas and the vibrational temperature of molecular species can be independently obtained.

The present exploratory study was made in the Re-entry Nose Tip (RENT) Facility at the Flight Dynamics Laboratory, Wright-Patterson Air Force Base in order to determine the feasibility of applying the LRS technique to an Arc-tunnel flow calibration.[1]

II. ARC-TUNNEL RAMAN SPECTROSCOPY

Arc-heated wind tunnels are used to generate high Mach number flows at high enthalpy. To accomplish this, the gas in the stagnation region must be heated to temperatures at which air constituents begin to dissociate. The gas is then expanded in a convergent-divergent nozzle where chemical recombination and vibrational relaxation occur, often at non-equilibrium rates. Hence, the effluent at the nozzle exit can be at a vibrational temperature which is greater than or equal to the high static temperature and could contain many chemical constituents such as N_2, O_2, NO, O, N, and NO^+ in varying concentrations depending upon the particular run conditions.

When aerodynamic bodies are placed in the test region (i.e., free-stream) of such tunnels, their forward region tends to stagnate the flow. Thus, conditions which were prevalent in the nozzle stagnation region (i.e., high temperatures and pressures) are again produced. The flow then expands around the bodies and free-stream conditions are again approached in the wakes of these aerodynamic models. Hence, some interesting thermochemical regions exist in arc-tunnel flows which should stimulate and challenge many Raman spectroscopists.

Along with all the interesting aspects of the arc-tunnel environment, one must also realize the problems that exist which hinder spectroscopic analysis in these facilities. The arc, which is used to heat the test gas, requires high electrical currents and voltages. Fluctuations in these quantities induce electrostatic and magnetic fields which can affect electronic instruments and electric signals. Also, these fields may critically affect the photomultiplier tubes used in spectroscopic data acquisition. Hence, the electrical environment of all arc tunnels can hinder spectral measurements. To minimize these possible induced field effects, electrostatic and magnetic shielding of electronic instruments should be used, especially if they are in close proximity to the arc. Also, double shielded signal cables will reduce some of these unwanted effects. Finally, a mu-metal shield for the photomultiplier tube is recommended.

The high effluent and electrode temperatures, which are common to arc tunnels, also create several possible light absorption problems. The chemical constituents of high temperature air include NO_2 and NO. Due to erosion and evaporation of the electrodes in arc-heated facilities, metal particle and metal vapor absorption can also occur. This type of absorption is broad-band (i.e., it covers a large portion of the spectrum), so nothing can be done to avoid it. If absorption is not negligible, the spectroscopic data must be corrected to account for this adverse effect. It is possible for electrode material absorption and/or molecular absorption to nearly obscure the spectral data sought; therefore, the experiment should be designed to minimize and/or correct for possible absorption.

Due to the critical arc-heater cooling requirements and high electrical power consumption, run times may be quite short in typical arc-tunnel facilities. This presents a practical problem of obtaining the desired spectral intensity versus wavelength scans if the photo-electric signal is weak, and some time must be spent gathering data at any given wavelength.

Another problem inherent in some arc facilities is acoustic noise and its corresponding deleterious vibration effects on experimental apparatus. The vibration of the optical components as well as electronic instruments can present serious problems to the spectroscopist. Acoustical insulation and vibration mounting must be employed in the design of the spectroscopic measuring device if acoustic noise is a problem.

One of the most difficult problems in arc-flow diagnostics is that of electromagnetic radiation coming from the hot flow. This emission can obscure the spectral signal of interest, especially if the signal is relatively weak.

One source of flow radiation is chemiluminescence which comes from the chemical reaction of NO with O to form NO_2. An excited molecular complex is assumed to be created in the reaction process which results in a continuum radiation being emitted during the completion of the reaction. A plot of radiation intensity versus wavelength which was obtained by Mastrup[2] is shown in Figure 1. This continuum flow emission covers a wide range of the spectrum, from blue to near-infrared, and is, therefore, difficult to avoid. Also shown in Figure 1 are the laser exciting line and Raman Stokes and anti-Stokes spectral regions for the nitrogen, argon and ruby laser sources (represented by the

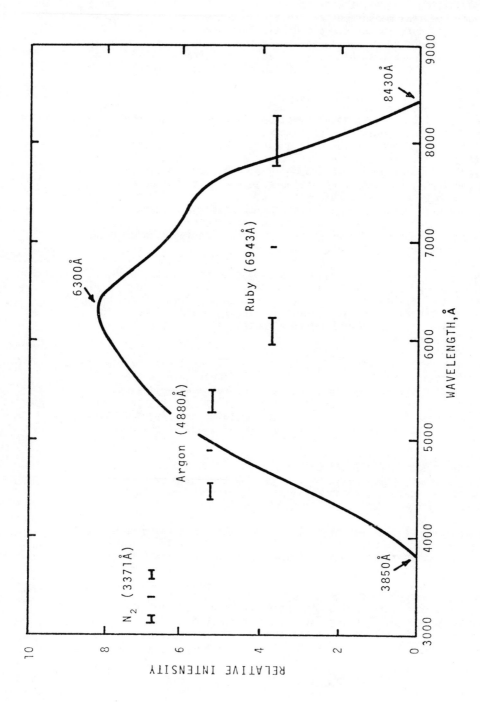

Figure 1. Measured continuum flow radiation attributed to NO + O chemiluminescence.

horizontal lines). It is seen that the choice of the ultraviolet pulsed
nitrogen laser for LRS - Arc-tunnel applications would completely
avoid the chemiluminescence problem, while Raman scattering experiments
utilizing the ruby and argon ion laser would have to contend with this
background flow radiation. Other sources of unwanted radiation from the
arc-heated effluent are the NO band radiation at wavelengths less than
3000 Å and the scattering of arc-radiation by electrode material in the
flow. The scattered light originates in the stagnation region and passes
through the nozzle throat. This so-called "flashlight effect" extends
from about 3200 Å to the near infrared and consists of molecular band
radiation, free-free and free-bound continuum radiation and atomic line
emissions. The particle scattered radiation of the "flashlight" effect
cannot be avoided through the choice of laser so it must be overcome
by some other approach, if it is not negligible.

Another important source of background radiation comes from
spontaneous emission by excited atoms of copper and silver electrode
material which are present as an impurity in the flow. Estimates
of the relative intensity of various copper emission lines[2] and a
silver doublet are shown in Figure 2. The strongest lines are located

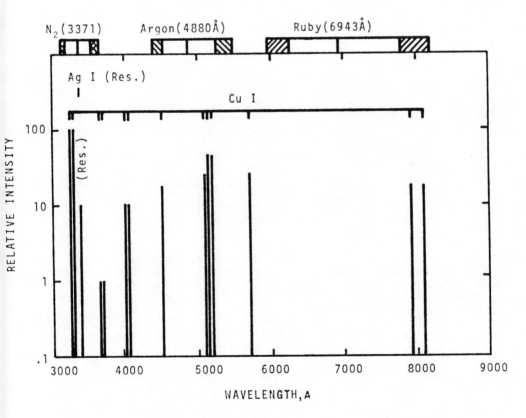

Figure 2. Estimated atomic line radiation from copper and silver.

near 3200 Å and 5000 Å. The 3200 Å doublet corresponds to the copper
resonance lines. The nitrogen, argon and ruby laser exciting wavelengths and Raman regions are also plotted in Figure 2. Note that
only the ruby Stokes region is overlapped by strong copper lines. Since
these emissions result from atomic transitions, they occupy narrow regions
of the spectrum and are usually not difficult to avoid. However, several
exceptions exist. The resonance lines can be pressure broadened, which
will widen their spectral region of influence. Also, intense radiation
from these lines, or any other source for that matter, can show up as
scattered light inside the spectroscopic measuring equipment, thus
presenting a problem.

The pressure and temperature environment, as well as the arc power
of each arc-tunnel, will dictate which of the above radiation sources
will dominate the emission spectrum of a particular facility. Thus, a
prerequisite to obtaining particular spectroscopic data should be an
investigation of the flow emission spectrum. An obvious advantage
of Raman spectroscopy is that the choice of a laser source can be made
so as to avoid the worst background noise problem areas. Chopped CW
or pulsed lasers of narrow pulse half-width should be used in conjunction
with lock-in or gating techniques in order to recover the signal from
flow background emission. In other words, experimental ingenuity
must be used to solve flow radiation problems.

III. RENT RAMAN INSTRUMENTATION

A schematic diagram of the LRS instrumentation used for the present
arc-tunnel measurements is shown in Figure 3. The light source was a
Model C102 Avco pulsed nitrogen laser, which produced 10-nanosecond half-
width, 100-kilowatt pulses at a variable rate up to 100 pulses per
second and delivered a maximum average power of 0.1 watt. The unpolarized
ultraviolet laser output was at a wavelength of 3371 Å with an output
bandwidth of 1 Å. Figures 1 and 2 show that the choice of the N_2 ultraviolet laser largely avoided regions of strong flow background emission
in arc-heated air flows. Furthermore, in this wavelength region no
significant absorption of incident or Raman radiation was expected
by the gas flow constituents, except possibly by NO_2 in the mixing
regions at the edges of the tunnel hot air jet.

A filter was used in front of the laser to block unwanted radiation
from the N_2 spontaneous emission in order to obtain a more nearly
monochromatic source. The transmission of the filter was 61 percent
at the laser frequency and 0.05 percent at 3577 Å, which is the location
of the strongest emission band.

The laser had to be located 26 feet from and downstream of the
measurement point in order not to interfere with the high priority
RENT tests. Hence, a somewhat complex and non-optimum "transfer and
focus" optical system had to be employed. The system consisted of a

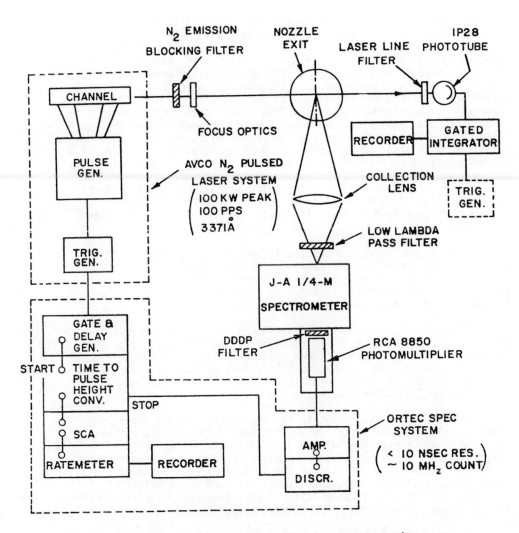

Figure 3.   Block diagram of experimental instrumentation
utilizing single photoelectron counting (SPEC) system.

flat mirror which reflected the laser beam into a 6-foot focal length
spherical mirror. The focused light from the spherical mirror gave an
image size of about 2 mm x 1.3 cm at the tunnel centerline just downstream
of the nozzle exit.

The intensity of the laser source was monitored by an RCA 1P28
Phototube after the beam had passed through the tunnel flow region in
order to determine if any of the incident beam radiation had been
absorbed by the test gas. A narrow band-pass filter, centered at
3371 Å and located in front of the 1P28 phototube, insured that
affects of flow radiation would be reduced. The phototube output
pulses were fed to a PAR Model 160 Box Car Integrator, which was
triggered by synchronous pulses from the laser trigger generator.
Integrator gate times of 30 nanoseconds were employed to provide a
laser beam pulse height analysis while rejecting steady-state background
radiation by a factor of 3×10^5. The analog output of the gated
integrator was then plotted versus time during a tunnel run. Linearity
of the system was checked by inserting various neutral density filters
into the beam path and recording the integrator output voltage change.
The elimination of flow radiation affects was confirmed through
measurements made by the monitoring system before, during, and after a
tunnel run.

Raman photons scattered vertically upwards were reflected 90° by
a mirror which, based on RENT facility non-interference considerations,
was positioned on an I-beam 58 inches above the test section.   The
photons were then collected by a 4-inch diameter f/23 collector lens
located 92 inches from the scattering volume.

A low wavelength pass filter was placed in front of the spectrometer
slit, whose purpose was to block radiation of wavelengths higher than
the Raman region in order to reduce scattered light problems arising
inside the spectrometer. A second filter labeled DDDP in Figure 3 was
employed in front of the photomultiplier tube to block scattered light
at, and below, the laser wavelength.[3]

The spectral analyzer used in the present experiment was a 1/4-
Meter Jarrell-Ash Ebert Scanning Spectrometer. This instrument was
operated in the first order with 500 μm slits and a 1180 grooves/mm
grating blazed at 3000 Å, giving a spectral bandpass of approximately
19 Å. The f/3.5 aperture ratio of the instrument provided good light-
gathering capabilities. Also, the small size and light weight of
the spectrometer made it quite suitable as a practical measuring device
for the present wind-tunnel diagnostics study.

The basic Jarrell-Ash instrument was altered to enable remote
selection of several fixed wavelengths within the Raman spectral region.
This was accomplished by placing mechanical stops on a pulley attached
to the wavelength drive screw of the spectrometer. Remote rotation of
the wavelength drive was then accomplished with a Selsyn transmitter-
receiver combination. The stops contacted either side of a retractable
plunger, which provided a fixed wavelength choice for both clockwise and

counterclockwise pulley rotation. Remote retraction of the plunger by a linear solenoid enabled other pairs of stops to be used, so that more than two known wavelengths could be remotely selected by using the "Selsyn" in conjunction with the solenoid.

An RCA 8850 Quantacon Photomultiplier Tube was employed to convert the spectrally analyzed photon flux into a photo-electric pulse. The 8850 PMT was selected because of its high gain and expected excellent quantum efficiency of 30 percent in the Raman Stokes wavelength region. The tube was operated at 1950 volts.

The pulse analysis system shown in the block diagram of Figure 3 consisted of a combination of off-the-shelf ORTEC Nuclear Instrumentation Modules (NIM) and was designated as the SPEC (Single Photoelectron Counting) System in Figure 3. The purpose of the system was to recognize and count laser Raman scattered photoelectrons from among those produced by PMT thermionic emission or continuous flow background radiation. The SPEC electronics system was quite similar to ordinary photon counting instrumentation except that it was designed for use with fast pulsed lasers. Here, "fast" is used in the sense that existing photon counting equipment could not distinguish multiple photoelectric events during the short laser pulse lifetime. Hence, the SPEC system works on what might be called the 1-0 principle. That is, each incident laser pulse must produce either one photoelectron event or none at all at the PMT cathode. The 1-0 principle is applicable only in the low Raman intensity regime, which was expected in the arc-tunnel experiments. The single photoelectron counting electronics were chosen to provide gating capabilities (i. e., pulse resolution) of the order of 10 nanoseconds while handling an input count rate of up to 20 MHz. Hence, the system was capable of reducing the dark current and flow background emission by a factor of 10^6.

In a pre-RENT test trial, the ability of the SPEC electronics to drastically reduce the continuous background radiation was checked by exposing the PMT to various arbitrary light signals of about the same intensity as expected from flow radiation during the arc-tunnel tests. The trigger generator was pulsed at 100 Hz, and a 100-nanosecond gate time was employed, which implied a noise reduction factor of 10^5. The anode current was measured with a picoammeter. The anode count rate and the rate meter reduced count rate were also recorded. The expected reduction of background signal was achieved, and the electronic gating method was therefore validated.

IV.   RESULTS AND DISCUSSION

Some of the difficulties in making spectroscopic measurements in arc-heated flows are discussed in Section II. Of those mentioned, background flow emission and incident laser beam absorption proved to be the major problems. Electrostatic, magnetic and acoustic noise was

eliminated by various shielding methods. Double shielded cables and a
mu-metal PMT shield proved most effective in overcoming the electrostatic and magnetic noise. Enclosing the spectrometer in a plywood
box lined with acoustical insulation eliminated the acoustic vibration
problem. Also, short tunnel-run-time problems were circumvented by using
the rapid wavelength selection method, involving the rotary-drive
mechanical stop system. Hence, adequate time was available to obtain
the basic spectroscopic data.

Figure 4 shows the results obtained from the flow emission and laser
beam absorption measurements. Note that the attenuation was directly

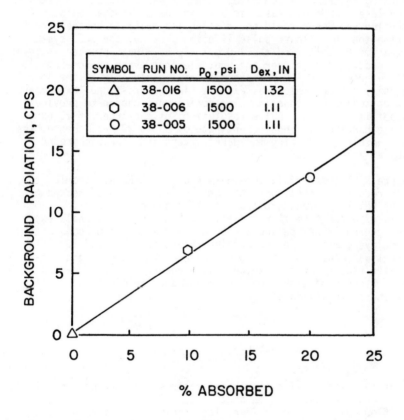

Figure 4. Laser beam attenuation and flow emission for various
           arc-tunnel runs.

proportional to the flow radiation. Hence, a common source for both
phenomena was suspected. Physical evidence of copper pitting of models
during the runs with highest emission and absorption leads one to suspect
that copper particles and vapor were the causes of the emission and
absorption problems encountered in this particular arc-tunnel facility.
This observation agrees with rapid scan emission data taken in the RENT
facility by Lawrence.[4] Note that the count rates in Figure 4 are not
actual, but rather effective values. That is, the true background count
has been reduced by a factor of 10^6 by the pulsed laser-gated photon
counting system. It is this effective rate which will tend to mask the
Raman signal.

The highest attenuation and flow radiation were measured for a
tunnel run which was unsuitable for model testing. For the usable runs
the laser attenuation was less than 10 percent and the background count
rate was less than 7 cps.

After concluding the above investigations, Raman scattering measurements were attempted for nitrogen in the arc-heated air plasma. A three-
point measurement was made with a broad (19Å 1/2 BW) apparatus function.
Scattering measurements were obtained at a wavelength near the (0, 1)
band peak and also in the "blue shifted" region of the Raman profile.
Also, a background count measurement was made ahead of the Raman region.
The results of the Raman measurements are shown in Table I along with

Table I

Comparison of Measurements with Theory

Quantity	Calculation	Raman Measurement
$T_{v_{N_2}}$	3570°K	<600°K
n_{N_2}	2.7×10^{19} cm^{-3}	$0.65 \times 10^{19} \pm 15\%$

the theoretically expected values of number density and temperature.
The comparison of theory with experiment indicates that the measured
temperature was much lower than expected from equilibrium calculations
performed by the Flight Dynamics Laboratory. Also, the number density
was one-fourth the theoretical value.

A plausible explanation of the poor comparison of theory with experiment for temperature data may be that the actual measured values were spatial averages of a large portion of the radial profile rather than point measurements of centerline values, as assumed in the theoretical calculations. This explanation does not, however, apply to the number density discrepancies since this parameter should be nearly constant across the jet for the particular run conditions of the present tests. Perhaps mechanical vibration, which could cause misalignment of the receiver optics and the incident beam path, was responsible for the lower measured signal. Of course, vibration would not affect the temperature measurements since only relative intensities are required. Finally, the above measurements were made with the facility operating in the peaked enthalpy profile mode. This condition further compounded the spatial resolution problem since only a small region of the jet possessed the theoretically expected values.

For further details concerning various aspects of this study, the reader should consult Ref. 1.

V. CONCLUSIONS

The experiments described in this report demonstrate that it is feasible to make Raman scattering measurements in high pressure, high enthalpy, arc-tunnel flows. The results also suggest that measurements in low pressure facilities are possible. Hence, the experimental measurement of chemical and thermodynamic properties of the major molecular constituents in arc-heated flows, some of which were previously unattainable by direct means, should now be possible. This flow calibration method could also be used to evaluate the degree to which arc-facilities simulate free-flight conditions.

Experimental measurements revealed that deleterious effects of acoustic and electromagnetic noise were present and could be overcome by isolating the measurement system from the arc-tunnel noise sources. Laser beam attenuation, which was shown in this experiment to be due to copper particles in the flow and copper vapor absorption, was a minor problem but was not negligible for some tunnel runs. The major and continuous experimental problem was found to be flow background radiation which tended to mask the Raman signal. For the particular facility employed in this study, copper line radiation was identified as the dominant radiation source. The adverse effects of the above background noise were, however, adequately reduced by the Single-Photoelectron-Counting (SPEC) technique used in the data-acquisition system.

Comparison of measurements of the vibrational temperature of nitrogen at the nozzle exit with equilibrium flow theory showed that the radially averaged temperature data obtained in this experiment could not be compared with a theory that predicts centerline values. Also, the discrepancies between theoretical and experimental number density noted in this study pointed out a need to study mechanical vibration effects on the instrumentation. Therefore, further **measure**ments with an improved point measurement system must be obtained before quantitative comparison of theory and experiment can be made.

REFERENCES

1. A. A. Boiarski and F. L. Daum, "An Application of Laser-Raman-Spectroscopy to Thermochemical Measurements in an Arc-Heated Wind Tunnel Flow," Aerospace Research Laboratories Report No. ARL 72-0126 (1972).

2. F. N. Mastrup, "Development of Spectroscopic and Optical Scattering Diagnostics for Nonequilibrium Reacting Gas Flows," AFFDL TR 70-17 (1970).

3. D. A. Leonard, "Feasibility Study of Remote Monitoring of Gas Pollutant Emissions by Raman Spectroscopy," AVCO Research Report 362 (1970).

4. L. R. Lawrence, Jr., R. E. Walterick, T. M. Weeks, and J. P. Doyle, Jr., "Total Enthalpy Measurements from Blunt-Body Gas Cap Emission in Arc-Heated Wind Tunnels," AIAA Paper No. 72-1021 (1972).

BOIARSKI DISCUSSION

HENDRA - Why can't you use infrared absorption or emission measurements at the temperatures of interest here?

BOIARSKI - We were interested in obtaining information on the homonuclear molecules such as nitrogen and oxygen. Absorption, emission, and optically-induced fluorescence measurements have been attempted in arc-heated wind tunnels for molecules such as NO. However, the background radiation and absorption from residual amounts of water vapor have caused problems. Therefore, little useful data has been obtained using these techniques. The only other method which has been successful in the arc environment is the electron beam-induced fluorescence technique, but this method cannot be applied to high-density flows.

I should mention that a detailed report concerning the application of LRS to arc-tunnel diagnostics is available. If anyone is interested, they should contact either Fred Daum or me.

RESULTS OF RECENT APPLICATIONS OF RAMAN SCATTERING
TO RESEARCH PROBLEMS AT NASA LANGLEY RESEARCH CENTER

by

W. W. Hunter, Jr.

NASA Langley Research Center

Results presented at the Raman Workshop illustrated work in progress at the Langley Research Center to utilize Raman scattering as a flow field diagnostic tool with emphasis on fluid mechanics and aerodynamics research applications. Results of two experiments conducted by Mr. M. E. Hillard were shown.

The first case illustrated the potential of the Raman technique for obtaining gas density measurements at a point in a hypersonic flow without disturbing the flow field. In this case a density profile across a normal shock on a hemisphere in CF_4 gas was obtained. See Reference 1 for details and results.

The second case illustrated the capability of the Raman technique to discriminate between various gas species and the method by which this advantage was used in a fluid piston experiment. Basically the problem was to determine the time for a driver gas to displace a driven gas. Details and results of the work is discussed in Reference 2.

---

1. M. E. Hillard, Jr., M. L. Emory, and A. R. Bandy, "Remote Sensing of CF4 Number Density in a Hypersonic Flow Using Raman Scattering," AIAA J. 11, 775 (1973).

2. M. E. Hillard, Jr., R. W. Guy, and M. L. Emory, "Raman Scattering Applied to Gas Mixing in the Successive Channel Flow of Two Gases," in Proceedings of International Congress on Instrumentation in Aerospace Simulation Facilities, California Institute of Technology, Sept. 10-12, 1973.

HUNTER DISCUSSION

DAIBER - Have you tried Raman scattering in a wind tunnel?

HUNTER - In a CF_4 facility. As a matter of fact, we performed these tests in a pilot facility and, based on these results, we are now building a detector system using interference filters. Also, you can extract temperatures from the ratio of intensities of the Stokes and anti-Stokes bands.

BERSHADER - Were you measuring the strength of the shock wave in that CF_4 facility?

HUNTER - We measured the density ratio across the shock.

BERSHADER - What was the percent uncertainty in that measurement?

HUNTER - The free stream measurements were typically accurate to 7 percent.

GOULARD - Is it an absolute measure of the density? That is, do you exactly know the volume?

HUNTER - Right, it is an absolute density measurement. You can approximate the sample volume from the geometry of the optics. If you do a calibration for ambient conditions, you eliminate the density-independent functions, but, of course you have to consider the temperature effects. The static temperature for these tests was about $5°K$.

PENNEY - I was wondering about the diffraction effects introduced by turbulence. Could these obscure the definition of your scattering volume?

HUNTER - We haven't investigated that point so far.

BERSHADER - If you are working on a well-designed wind tunnel, the turbulence-induced density fluctuations can be very low, typically a few hundredths of a percent. Of course, if you are in a mixing region, that is something else.

May I comment on the density measurement methodology, because this is very interesting to me. I think there are cases where some of the standard methods of measuring density can be further exploited than they have so far. We have been hearing about some of the dipole interactions around resonance today and if one looks at the refractive index formulas around resonance, one finds peaks of refractivity which come at the half power points of a typical resonance curve. And n-1 which is the appropriate quantity, can go up, based on "back-of-the-envelope" calculations, by a factor of 10^5; thus, if in fact one has a dye laser that was tuned to the half power point (let's say you were using sodium vapor just as an illustration) you could see little vortices, or other small effects that would not be visible with ordinary interferometry or shadow methods. I think that this idea becomes an interesting possibility now that we have tunable dye lasers.

The laser is needed because of the fact that if you shine an ordinary sodium lamp on sodium vapor, then you don't know what you have, because the dispersion around the sodium line is so strong that you can't even use the standard group refractive index formulas. Brillouin has looked at this problem a bit, and it is a tough one.

Apart from resonance refractivity, the refractive methods for separating out component densities (or gradients) of a mixture is not particularly sensitive. In this connection one might well ask "why can't I do interferometry with two different wavelengths and get two equations and two unknowns and solve for my two densities?" The trouble is, with heavy particle species, unless there is a large difference in dispersion between the two species, the precision of that particular analysis will not be very good. Now when you have free electrons and heavy particles, the dispersion of the electrons is completely different from the dispersion of the heavy particle, and the method is fine. You can do two-wavelength interferometry if you have high enough electron densities and this is well known, as some of the people right here have shown in the early days. You get fine accuracy with this kind of technique. But in some of these mixtures of different species, e.g. in jets or flames, it is less satisfactory, so I think that the use of Raman scattering to measure partial concentrations of mixtures should be very worthwhile.

RAMAN SCATTERING MEASUREMENTS OF MEAN VALUES
AND FLUCTUATIONS IN FLUID MECHANICS

by

Samuel Lederman

Polytechnic Institute of New York Graduate Center
Farmingdale, New York

ABSTRACT

The purpose of this paper is to show that Raman scattering excited
by a pulsed laser can be used to obtain useful information about both
the mean values of gas properties and their fluctuations in a dynamic
system. This conclusion is illustrated by data from measurements on a
CO_2-air jet and a CO_2-enriched flame.

* * * * * * * * * * * * * *

As was pointed out by a number of speakers at this meeting, it is
fairly well-established by now that Raman scattered intensities are proportional to the concentration densities of the observed species, and that
use of the ratio of some combination of rotational line intensities, or
some ratio of the vibrational Stokes intensities of different vibrational
levels, will provide the temperature. Most, if not all, reported experimental data were obtained by photon counting and scanning techniques. The
time to obtain a measurement ranges from a fraction of a second using the
techniques of Prof. Delhaye or Dr. Smith, as described elsewhere in this
Proceedings, to several tens of minutes by the more conventional techniques,
and in general an average or mean value at a given point in space is
obtained.

At this time I want to illustrate that "instantaneous" measurements
can be obtained using a single pulse technique. As an example in point,
the system shown in Figure 1 was used to determine the instantaneous
concentration of a specie and, with minor modification of the system, the
instantaneous temperature in a flame. The modification for the temperature measurement consisted of substituting a CO_2 enriched flame for the
axisymmetric turbulent jet of air and CO_2. Some of the data are shown in
Figures 2 and 3.

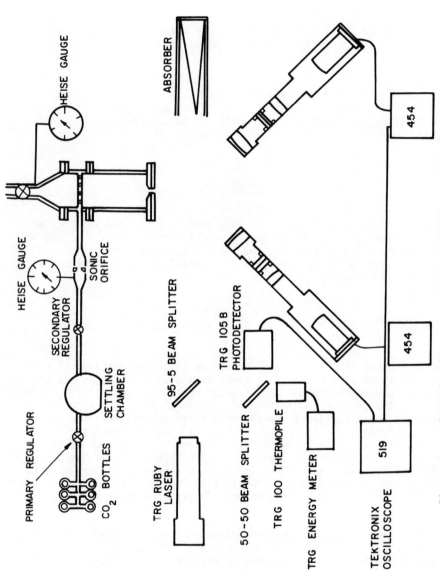

Figure 1. Schematic diagram of the concentration measurement system.

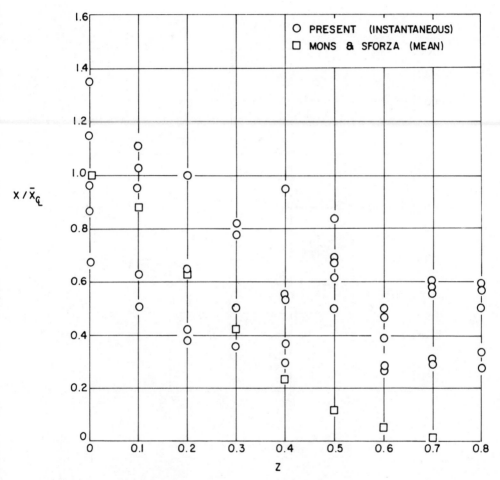

Figure 2.   Instantaneous specie concentration of CO_2 in a turbulent
jet at $x/D = 10$.

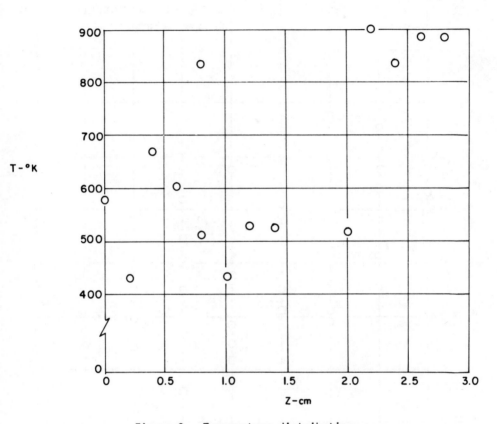

Figure 3.   Temperature distribution.

As can be seen from the figures, the data are widely scattered. However, the expected statistical fluctuation in photon counts is only 5-10% for this system. What, then is the reason for this apparent excessive scatter?

To answer this question, we constructed a detection system with a repetitive laser source (6 pulses per minute), which was coupled to a data acquisition and reduction system. (See Figure 4.) At each geometric point, 50 Raman scattering data points were taken. While perhaps not adequate for complete statistical data analysis, this approach is considered to be sufficient for the proper interpretation of our results, as given in Figure 5. Here, we see that mean data values are obtained, and we have therefore established that most of the scatter results from fluctuations in gas properties within the measurement volume. In Figure 5, we also show mean values of the concentration for the same jet determined by a chromatographic measurement technique.

Thus, we have shown that useful mean values of gas properties can be obtained directly through use of multipulse laser sources, and that single pulse sources provide data on time-related fluctuations which, by proper averaging, permit the mean temporal values to be obtained.

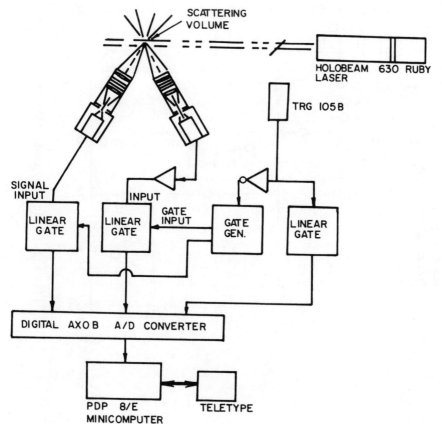

Figure 4.   Schematic diagram of the experimental configuration.

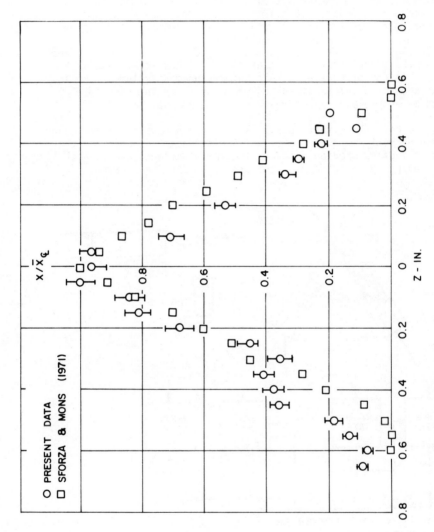

Figure 5. Specie concentration profile in turbulent jet at x/D = 5.0.

LEDERMAN DISCUSSION

LAPP - For each ruby laser pulse you must get a large number of photons. Earlier today Dr. Schildkraut, I believe, commented that a photon may give you signals of varying heights, except with a new RCA tube which had fairly small (10%) uncertainty in height, so that one could use multiphoton counting.

SCHILDKRAUT - It depends on how many photons you get back per unit time. They can bunch up and be non-resolvable.

LAPP - Right. I was wondering how Prof. Lederman's technique compares to straight photon counting.

LEDERMAN - Well, I used originally a static calibration system, where the peak amplitude as recorded on a scope was related to the number density. This calibration resulted in a linear relation between amplitude and specie concentration. The same sort of procedure was followed with the new data acquisition system. The signal was stretched according to the maximum amplitude across the phototube load resistor, digitized by an analog to digital converter, normalized with respect to the primary laser pulse and processed by the on-line digital computer. Saturation was avoided by proper scaling.

SCHILDKRAUT - The system we're discussing is not the same as photon counting and does not have quite the same low noise characteristics.

GOULARD - On one of your slides (see Fig. 5) you showed chromatography and Raman results, and in some cases they are quite far apart. Do you have any opinion why?

LEDERMAN - Well, again, we only used 50 points per point to obtain the mean values. An increase in the number of the averaged points will reduce the discrepancy between the chromatographic results and the Raman results. One must remember that with the chromatograph the mean value is obtained by sitting on the same point for 15-20 minutes. In any case the data as shown here are intended only to make the point that the apparent scatter is due to the turbulent nature of the flow field and not to the Raman single pulse measurement, which is instantaneous. By using photon counting the turbulent fluctuations might be obscured.

BERSHADER - Would you describe the flame used for the temperature measurements?

LEDERMAN - It was a methane flame with the addition of CO_2 for the temperature measurement. The Stokes and anti-Stokes lines of the CO_2 were measured. As far as the other jet was concerned, it was an axisymmetric air and CO_2 jet at subsonic speeds.

SMITH - How do you establish that you have a flame in local thermodynamic equilibrium, in order to utilize the Stokes/anti-Stokes measurement in CO_2?

LEDERMAN - Well, I did not establish anything; I just measured data--specifically, the Stokes and anti-Stokes intensities.

SMITH - Did the resulting temperatures agree with other measurements?

LEDERMAN - Yes. They did agree with results from thermocouples.

RAMAN GAS MIXING MEASUREMENTS AND RAMANOGRAPHY

by

Danny L. Hartley

Sandia Laboratories
Livermore, California

Application of Raman scattering as a gas dynamics diagnostic tool has been under development at Sandia since 1969. The early stages of this effort were directed toward developing a capability to measure transient mixing phenomena over a 1 second time interval. Transient concentration measurements of nitrogen injected into helium were obtained using an AVCO 100 pps N_2 laser[1]. More recent Raman gas mixing studies were aimed toward examining the effect of pressure level on mixing of H_2, D_2, N_2, and He[2]. These experiments collected Raman scattering from a 1 mm sample volume inside a closed vessel. An on-going effort to measure Raman scattering from steady state turbulent flows is shown in Figure 1. In this effort, direct measurement of Reynold's stresses are being attempted by P. O. Witze at Sandia.

Since the experiments conducted earlier were limited to single point detection, an experiment was devised and performed whereby an entire 2-D flow field could be concentration-mapped in one pulse. This technique, which I call Ramanography, is depicted in Figure 2. A thin sheet (1 mm x 75 mm) of laser light from a frequency-doubled ruby laser is passed through the non-homogeneous field created by interacting jets of one gas injected into a dissimilar gas. Raman scattered light from the injectant gas is collected, filtered by appropriate interference filters, intensified by a high gain image intensifier (10^6), and recorded on a photographic plate. The resultant image is a 2-D concentration map of the injectant gas in the plane of the laser sheet. Several such Ramanographs have been taken to date to show the versatility of such a technique.

Other applications of Raman scattering at Sandia are discussed here by my colleagues; J. R. Smith is measuring rotational temperatures in a pulsed system; R. A. Hill and I have developed a unique light trap for Raman work; and R. E. Setchell, C. Hartwig and I are extending our studies to turbulent reacting flows.

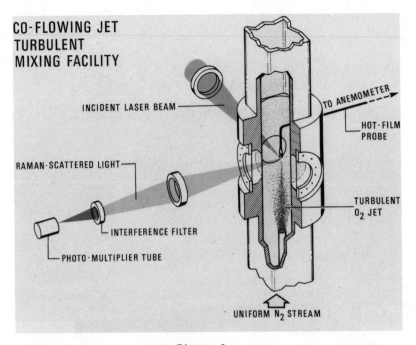

Figure 1.

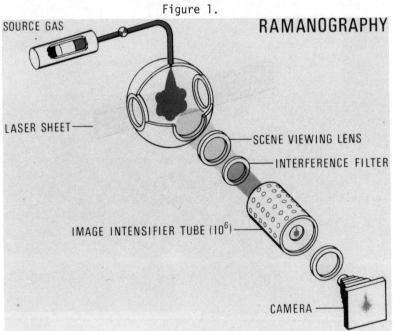

Figure 2.

REFERENCES

1. Hartley, D. L., "Transient Gas Concentration Measurements Utilizing Laser Raman Spectroscopy," AIAA Journal, Vol. 10, No. 5, p 687, May 1972.

2. Hartley, D. L., "Application of Laser Raman Spectroscopy to the Study of Factors that Influence Turbulent Gas Mixing Rates," Fluid Mechanics of Mixing, ASME, p 131, June 1973, also to be published in AIAA Journal.

HARTLEY DISCUSSION

WHITE - Has that word Ramanography been used in public before?

HARTLEY - I don't think so. It is my word.

WHITE - This may be a historic occasion!

DAIBER - Are you going to multi-pulse that laser and get sequential pictures of the mixing?

HARTLEY - I would like to. That would be another quantum step forward. Originally I wanted to do that, but the pictures I have now were taken a millisecond apart. It is a very fast process that I'm looking at. I look at a jet come in and try to take a fast enough exposure in order to freeze the eddies. This is easily accomplished with a 10 nanosecond laser pulse. I then rerun the experiment, change my timer so that it is one millisecond later, take another picture, and so forth. And when you are running it at several hundred psi, which my pictures correspond to, it takes a long time to pump that vessel up.

DAIBER - Do you get good run-to-run reproducibility so that you can get a map that way?

HARTLEY - Surprisingly so, in that the global properties, such as jet size, front location, and large eddy character are very reproducible. I can stop the jet at, say 5 milliseconds from the time I open the valve. I can rerun it at 5 milliseconds again and the jets are in the same place, although some of the fine structure certainly is different.

BERSHADER - Are you tilting the interference filters to get spectral resolution?

HARTLEY - It isn't necessary to tilt them. My filters are selected to pass the first Stokes vibrational Raman wavelength of my injectant gas with a band pass of 50 Å.

BERSHADER - There is one point regarding interference filters that I would like to mention here. Using interference filters for spectral resolution, we found that the transmission wavelength - in particular, the bandwidth as a function of angle of tilt of the filter - was very sensitive to the polarization of the light. So if you have any change of polarization, then there is a problem. In our work on electron scattering of pulsed laser light, we had to do a lot of calibration of interference filters since we couldn't find anything in the literature on the effect of different planes of polarization on the spectroscopic behavior of the thin film filters as a function of angle. I would appreciate hearing from anyone who knows more about this point.

RAMAN SCATTERING WITH HIGH GAIN MULTIPLE-PASS CELLS*

by

R. A. Hill
Sandia Laboratories
Albuquerque, New Mexico

D. L. Hartley
Sandia Laboratories
Livermore, California

Presented by R. A. Hill

ABSTRACT

We describe a simple optical system that utilizes a unique property
of ellipsoidal mirrors, viz., light brought to one focus will be
reflected alternately through the two foci and collapse to the major
axis. Raman-scattered light from atmospheric N_2 indicate that gains
on the order of 100 are achieved with this new system.

I. INTRODUCTION

Photoelectric recording of Raman-scattered light in gases where
the scattering cross sections are small requires that a relatively
large light flux be provided at the point of observation. This can
be accomplished by reflecting the incident light, say from a laser,
through the observation region a number of times. One such method, the
multiple-pass cell described by Weber, et. al.,[1] uses two plane mirrors
to allow the laser beam to traverse a finite volume several times.
As pointed out by Barrett and Adams,[2] this system has the disadvantage
that it is difficult to optically couple a large fraction of the Raman-
scattered light to the spectrometer. A second technique whereby the
sample is placed in the laser cavity has been used also to increase
the Raman scattering efficiency. Such methods, however, are limited

---

*This work was supported by the U. S. Atomic Energy Commission.

in gain because the laser cavity approaches an unstable mode of operation after a finite number of traverses through an inter-cavity cell.[1] A third technique, described by Barrett and Adams,[2] involves focusing the laser beam at the point of observation. This method has the advantages that there is no need to illuminate a large gas sample, and that all of the Raman light scattered into a large solid angle can be passed through the slit and aperture stop of the spectrometer.

This paper describes a new, particularly simple optical system that serves to reflect the incident laser light through the same point a large number of times with the laser beam focused at the point of observation. This system makes use of a unique property of ellipsoidal mirrors, viz., light brought to one focus of an ellipsoid will be reflected alternately through the two foci, and after a number of reflections, will approach parallelism with the major axis. This light-trapping concept,[3,4] is further explored in Section II. Two different versions of a practical "light-trapping cell" are discussed in Sections III and IV, and the results of an experimental check of the increase in Raman-scattered signal is presented in Section V.

II. LIGHT-TRAPPING CONCEPT

Consider an optical system consisting of two identical on-axis ellipsoids positioned such that their respective foci are coincident, as shown in Figure 1. If light is focused at f_2 such that it subsequently falls on mirror M_2, the light will be reflected by M_2 and focused at f_1. A light ray that strikes M_2 at a radius r thus strikes mirror M_1 at a radius $r(1 - e)/(1 + e)$ where $e = c/a$ is the eccentricity, $a = 1/2$ major axis, and $c = 1/2$ distance between foci. The light is subsequently reflected by M_1 and focused at f_2, and the light ray that originally struck M_2 at a radius r now strikes M_2 at a radius $r[(1 - e)/(1 + e)]^2$. Thus, the light is reflected alternately through the two foci and after a number of reflections, the light rays approach parallelism with the major axis.

The maximum angular aperture α of the system shown in Figure 1 can be found from $\tan \alpha_0$ and $\tan \alpha_1$, with the result

$$\tan \alpha = \frac{2a \, re \, (1 - e^2)}{a^2(1 - e^2)^2 - r^2 e^2} \tag{1}$$

Since the f/number, F, is given by $(2 \tan \alpha/2)^{-1}$,

$$F = a(1 - e^2)/2 \, re. \tag{2}$$

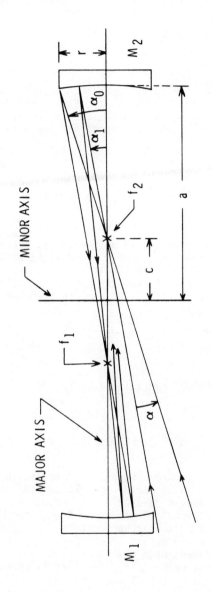

Figure 1. A schematic of a light-trapping system with identical on-axis ellipsoidal mirrors M_1 and M_2. Light incident at focus f_2 is reflected alternately through the two foci.

The minimum diameter d_0 of the first image at f_2 can be taken as the width of the central diffraction pattern due to a circular aperture. This is the diameter of the Airy circle d_0 = 2.44 λF where λ is the wavelength. Because of magnification, the subsequent image diameter at f_1 is $d_0 M$ where $M = (1 + e)/(1 - e)$ is the magnification, and the nth spot diameter is

$$d_n = d_0 M^n, \quad n = 0, 1, 2,... \tag{3}$$

While the spot diameters of the light passing through each successive focus are magnified, the spot diameters of the successive return beams in the "focal planes" are demagnified and these successive "spots" converge to the major axis as M^{-n}. When the magnified image diameter is larger than

$$D = 4c \tan \alpha_1 = 4cr \left\{ c + a\left(1 - \frac{r^2}{a^2 - c^2}\right)^{\frac{1}{2}} \right\}^{-1} \tag{4}$$

light escapes from the system in a cone with half-angle α_0 around the periphery of the ellipsoid as shown in Figure 2. The light with half-angle less than α_1 is intercepted by an ellipsoid and thus makes an additional pass before the outer part leaves the system, etc. The number of passes through the foci before light escapes from the system is, from Eq. (3),

$$n = INT \left\{ \ln(D/d_0)/\ln M \right\} \tag{5}$$

where INT $\{X\} \equiv$ integer obtained by truncating X.

III.   LIGHT-TRAPPING CELL (LTC I)

The optical system shown in Figure 1 serves to increase the light intensity at both foci; however, to be useful as a Raman-scattering cell a more appropriate system should have only one focus. A schematic of such a system that uses an ellipsoidal mirror in the manner discussed above is shown in Figure 3. This "light-trapping cell" (LTC I) consists of an on-axis ellipsoidal mirror and a flat mirror positioned such that its reflecting surface is coincident with the minor axis of the ellipsoid and its normal is parallel to the major axis. The flat mirror serves to reflect the rays that approach focus 2 back through focus 1. Thus, a large number of images of the initial image are focused at a single focus. As above, if the initial spot diameter in the "focal plane" is d_0, the

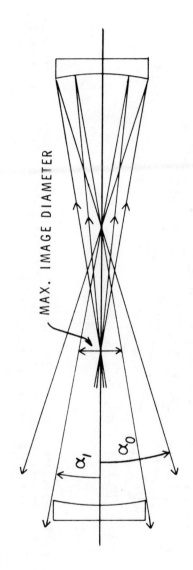

Figure 2. The maximum magnified image diameter is defined by the angle α_1; light eventually escapes from the system in a conical annulus defined by the angles α_0 and α_1.

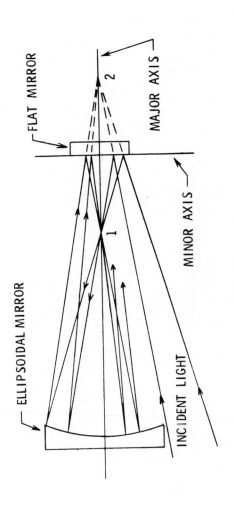

Figure 3.   Schematic of a "light-trapping cell" (LTC I). The flat mirror is positioned half-way between the foci and faces the ellipsoidal mirror. Light incident at focus #1 passes repeatedly through focus #1.

n^{th} spot diameter is given by Eq. (3). Thus, the successive spot diameters of the light passing from the flat through the focal planes are magnified while the successive spot diameters of the return light passing from the ellipsoid through the focal plane are demagnified and these successive spots converge on the major axis as M^{-n}. For a flat with diameter $D_f \geq 2c \tan \alpha_0$ (α_0 is defined in Figure 1), light eventually escapes from the system in a cone with half-angle α_0 around the periphery of the ellipsoid. The light with half-angle less than α_1 is intercepted by the ellipsoid and makes an additional pass to the flat before the outer part leaves the system, etc. A simple calculation of the increase in Raman scattering from the focal region of this system was carried out previously and will not be repeated here.[4] A modification of this system that results in gains approximately four times that with the above system is discussed in the following section.

IV. LIGHT-TRAPPING CELL (LTC II)

Consider an LTC with a flat mirror whose diameter is exactly one-half of the maximum image diameter, i.e., $D_f = 2c \tan \alpha_1$. For such a system, light escapes from the system in a cone with half-angle α_0 around the periphery of the flat. Light with half-angle less than α_1 is intercepted by the flat and makes an additional pass to the ellipsoid before the outer part leaves the system, etc. If this flat mirror is circumscribed with a spherical mirror whose origin is coincident with the near focus of the ellipsoid, i.e., whose radius of curvature is $c \sec \alpha_0$, the escaping light will be directed back into the LTC II along an inverted path. As shown in Figure 4, this coaxial spherical-flat mirror is pierced to provide access for the light entering the cell. Because the redirected light follows an inverted path, the light will make as many passes through the cell as in the first set of passes and eventually arrive at the sphere on the side of the flat mirror opposite the entrance hole. Thus, the light that made 2n + 2 passes through the focal region before arriving at the sphere makes a total of 4n + 4 passes before returning to the sphere at a spot opposite the entrance hole. The light that made 2n + 4 passes through the focal region before arriving at the sphere makes a total of 4n + 8 passes before returning to the sphere at the same spot opposite the entrance hole, etc. Thus, each successive outer cone of light that is redirected back into the cell by the spherical mirror eventually arrives back at the spherical mirror at a point diametrically across the flat mirror from the entrance hole. This light is returned through the origin of the spherical mirror (near focus of the ellipsoid) where it goes through a third set of passes, inverted with respect to, but otherwise identical in nature to the first set of passes. As in the first set of passes, light eventually tries to escape around the periphery of the flat mirror where it is returned by the spherical mirror for a fourth set of passes, inverted with respect to, but otherwise identical to the second set of passes. This light eventually returns to the sphere at the entrance hole where it escapes from the cell.

A simple calculation of the increase in Raman-scattered signal has been carried out for the case where the focal region is viewed with a

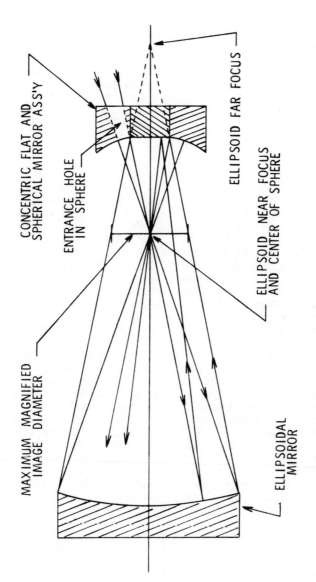

Figure 4.   Schematic of a "light-trapping cell" (LTC II) equipped with a coaxial flat-spherical mirror assembly. The spherical portion of the coaxial mirror serves to redirect escaping light back into the system.

spectrometer slit oriented perpendicular to the major axis.[5] This calculation assumes a slit length at least as large as the maximum image diameter D and, for ease of calculation, a uniform intensity distribution
for the initial image.

For all four sets of passes,

$$I = \frac{I_0}{I - R} \left\{ 1 - \frac{(M^2 - 1)^2 R^{8n + 8}}{(M^2 - R^4)^2} \right\} \qquad (6)$$

which represents the total intensity in the focal region of the LTC II
in terms of the intensity I_0 in the first pass. A plot of I/I_0 (the
gain) is presented in Figure 5 for a range of values of the eccentricity
and reflectivity. These calculations were carried out for a system
with a = 15.24 cm, r = 3.81 cm, and λ = 5000 Å. Figure 5 indicates
that high gains require both high reflectivity and low eccentricity.
For R = 1, Eq. (6) is indeterminate. Use of L'Hospital's Rule gives

$$I_{R = 1} = 8 I_0 \left(n + \frac{M^2}{M^2 - 1} \right) \qquad (7)$$

Also, as M approaches one, i.e., as the eccentricity approaches zero,
the gain approaches $I_0/(1 - R)$. However, as e → 0, the f/number becomes
large and the initial image diameter d_0 becomes equal to D at e ≈ 0.004.
Thus, as e → 0, n goes through a maximum and then goes to zero. As a
result, the gain has a maximum value at some small but non-zero value
of e. According to Eq. (6), for R = 0.99, a maximum gain of 99.99 is
obtained at e ≈ .009 and for R = 0.999, a maximum gain of 612 is obtained
at e ≈ 0.0085.

The dependence of the gain on the radius of the ellipsoidal mirror
at various values of the eccentricity is shown in Figure 6. These curves
were computed for R = 0.999 and a = 15.24 cm. According to Figure 6
larger diameter ellipsoids have a significant advantage only for e ≳ 0.15.

V.   TEST OF THE LTC II

An experimental LTC II has been constructed using an ellipsoidal
mirror with e = 0.2 and the dimensions listed above. Both the ellipsoid[6]
and the coaxial, flat-spherical mirror[7] assembly were dielectric coated
for 99.9% reflectivity in the wavelength range 4700 Å to 5200 Å. For
ease of focusing and alignment, both mirrors were held in mounts that
could be translated in a common slide assembly. A photograph of this

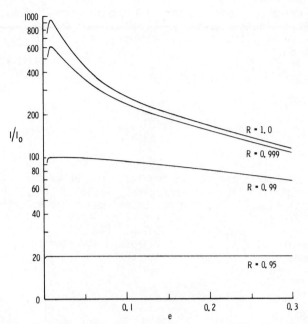

Figure 5.   The variation in calculated gain, I/I_0, with
            eccentricity for several values of reflectivity.
            These curves represent a system with a = 15.24
            cm, r = 3.81 cm, and λ = 5000 Å.

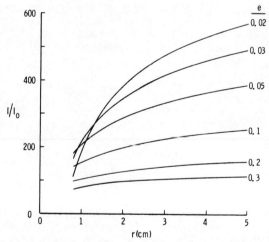

Figure 6.   The variation in calculated gain, I/I_0, with radius,
            r, of the ellipsoid for several values of the
            eccentricity. These curves represent a system with
            a = 15.24 cm, R = 99.9 percent, and λ = 5000 Å.

LTC II, illuminated with the 4880 Å line from an argon-ion laser, is
shown in Figure 7. The intensity increase at the focus of this cell is
very noticeable.

A Raman-scattering experiment was carried out to measure the intensity
obtained with the LTC II. The apparatus included a Spectra Physics 2-
watt argon-ion laser operating at 4880 Å, a Spex model 1800 spectrometer,
and an RCA photomultiplier detector (bialkali photocathode) that was
cooled to -20°C by a thermoelectric cooler. The Raman line corresponding
to the $\nu(1 - 0)$ vibrational transition (Stokes line) in atmospheric
nitrogen was recorded at 5506 Å; the Rayleigh background was recorded
at both 5400 Å and 5600 Å. The experimental gain was determined by the
ratio of the scattering with the cell to that obtained with one beam.
In order to compare the results of this experiment with that for the
LTC I described in Ref. 4, the height viewed was limited to 1.0 cm.
The measured gain of the LTC II was 93 ± 5, the range resulting from the
difficulty of measuring the Raman signal with only one beam. As
expected, this gain is approximately four times that obtained with the
LTC I. This result is somewhat fortuitous, because reflection losses
with the LTC II should limit the gain of the LTC II to somewhat less
than four times that obtained with a comparable LTC I. The difference
is probably in the focusing lens used in the LTC II. This lens was of
slightly shorter focal length, (5.0 cm compared to 5.1 cm) and was of
much higher optical quality with the result that the initial image
diameter d_o was somewhat smaller in the LTC II.

The pure rotational Raman spectrum of nitrogen was recorded by
placing a small, low-pressure nitrogen jet near the focal region of
the LTC II. This spectrum, reproduced in Figure 8, was recorded from
the output of a Kiethley picoammeter at a rate of 20 Å/min and shows
that Raman scattering can be easily carried out with the LTC II and a
conventional spectrometer.

VI. CONCLUSIONS

The maximum gain achievable with an LTC II is obtained for quite
small values of the eccentricity of the ellipsoidal mirror, say $e \simeq 0.01$.
In practice, unless a is sufficiently large, an LTC II with such small
eccentricity would be difficult to build and use. The flat mirror would
have a very small diameter and the scattering region would be extremely
close to the flat mirror. This could be circumvented by using two
ellipsoidal mirrors as shown in Figure 1, however, it would then be
necessary to back each ellipsoid by an appropriate spherical mirror to
redirect magnified escaping light back into the cell. The f/number of
such a cell would be quite large (~ 250); however, this would not present
a problem for a laser source. The most important parameter is the quality
of the reflecting surfaces. These surfaces must be kept clean, and a
reflectivity approaching 99.9% is necessary to obtain high gain.

Figure 7. Photograph of an LTC II, illuminated with the 4880 Å line from a 2-watt argon-
ion laser. The ellipsoidal mirror is on the near side; the coaxial mirror is
on the off side. The intensity increase at the focus is quite noticeable.

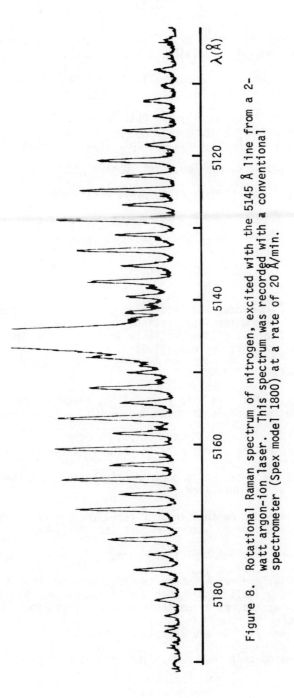

Figure 8. Rotational Raman spectrum of nitrogen, excited with the 5145 Å line from a 2-watt argon-ion laser. This spectrum was recorded with a conventional spectrometer (Spex model 1800) at a rate of 20 Å/min.

REFERENCES

1.   A. Weber, S. P. S. Porto, L. E. Cheesman, and J. J. Barrett, J. Opt. Soc. Am. <u>57</u>, 19 (1967).

2.   J. J. Barrett and N. I. Adams, III, J. Opt. Soc. Am. <u>58</u>, 311, (1968).

3.   R. A. Hill and D. L. Hartley, J. Opt. Soc. Am. <u>62</u>, 1359 (1972).

4.   D. L. Hartley and R. A. Hill, J. Appl. Phys. <u>43</u>, 4134 (1972).

5.   R. A. Hill and D. L. Hartley, Appl. Optics <u>13</u>, 186 (1974).

6.   Supplied by J. Unertl Optical Company, Pittsburgh, Pennsylvania.

7.   Supplied by Special Optics, Cedar Grove, New Jersey.

HILL DISCUSSION

BARRETT - If you focus the laser beam with a lens and position a spherical mirror beyond the focal region, so that the beam is returned back on itself, then all of the available photons in the laser beam will pass through the focal region many times.

HILL - Right.

BARRETT - It seems that with your system, since you have many reflections, you're probably going to end up with a larger sample size or sampling volume.

HILL - The sampling volume here is about 1 centimeter high.

BARRETT - I don't see how you can do any better with the amount of photons coming out of your laser.

HILL - It seems to me that in your case, the photons pass through the focus twice and then go back somewhere into the laser.

TARAN - Where they get amplified again.

HILL - But it has been my understanding, that in those kinds of systems, there is a limit to how much amplification you can get before the laser cavity approaches an unstable mode of operation.

BARRETT - Then you can use an optical isolator.

PENNEY - But doesn't that reduce the gain back to 2?

BARRETT - Perhaps so. For my experiments, I remove the laser output mirror so that my gas sample is inside the laser cavity.

HILL - But the idea here is that there are certain times when you can't do that. We certainly can't do it in applications such as gas mixing that is carried out in an external cell at very high pressure. Rather than passing unfocussed beams through an area between two parallel plates, this optical cell provides a well-defined focal region and has demonstrated high gain.

TARAN - I should like to comment on the arrangement with the Raman cell in the laser cavity. We made measurements in order to assess the improvement over one single pass of a 1 W argon-ion laser beam. We gained a factor of 30 in the scattered light intensity.

HILL - Well, we measured 92 for this cell; however, that is not quite fair. Perhaps we should compare it with sending the beam through the focus twice, which would mean we obtained a gain of about 50. But, according to the gain curve, for low eccentricity, the beam is not magnified as rapidly and thus stays in the cell much longer with resulting gains of several hundred.

PRELIMINARY RESULTS OF RAMAN SCATTERING
DIAGNOSTICS OF EXPANSION FLOWS

by

J. W. L. Lewis, ARO, Inc.

Arnold Engineering Development Center
Arnold Air Force Station, Tennessee

A hypersonic N_2 expansion flow field was produced using a conical nozzle of 1-mm throat diameter and exit area ratio of approximately 25. The investigation included N_2 reservoir pressures P_0 of 1 - 10 atm at 285°K, and the reservoir conditions were such that the axial region investigated included both isentropic and anisentropic, condensing N_2 flow. Raman scattering was utilized for measuring the N_2 gas density, and both axial and radial profiles were obtained. For the far-field, low-density region of the expansion the electron beam fluorescence technique was used to determine both the density and temperature of the N_2 expansion. Throughout the flow field Rayleigh scattering was employed to detect the onset of condensation and subsequent cluster growth. The depolarization ratio of the Rayleigh-scattered radiation was determined in order to observe the change in the scatterer's symmetry as clustering progressed.

The figure shows the results of the density measurement of N_2 as a function of axial distance for P_0 = 2.6 atm. These Raman data were acquired using a conventional ruby laser with approximately 36 joules/pulse. The method of characteristic solutions shows the predicted isentropic variation, and the far-field electron beam results are also shown. It should be noted that condensation existed for this condition for $x/D \gtrsim 15 - 20$ but with a negligible mass fraction of condensate. For a higher value of P_0 the depolarization ratio of the scattered light was observed to decrease from a value of approximately 8.5×10^{-3} near condensation onset, which is of the order of that of N_2, to approximately 1×10^{-3} following the cessation of cluster growth, thereby indicating the transition during growth

The research reported herein was conducted by the Arnold Engineering Development Center, Air Force Systems Command. Research results were obtained by personnel of ARO, Inc., contract operator at AEDC. Further reproduction is authorized to satisfy needs of the U. S. Government.

AXIAL VARIATION OF N_2 NUMBER DENSITY RATIO N/N_0

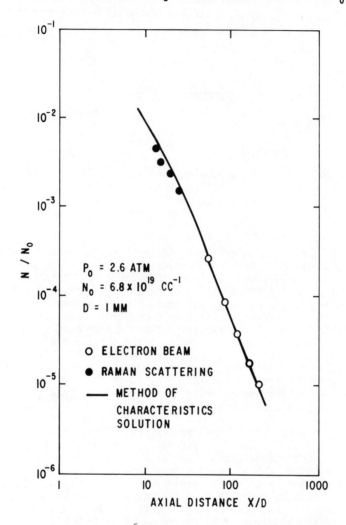

to a more symmetrical cluster scatterer. The experimental details
and analysis are to be found in Ref. 1, and work in progress includes
the measurement of gas rotational temperature and density using
rotational Raman scattering for N_2 and a variety of other species.

1.   C. C. Limbaugh, J. W. L. Lewis, M. Kinslow, H. M. Powell,
      L. L. Price and W. D. Williams, "Condensation of Nitrogen in a
      Hypersonic Nozzle Flow Field," AEDC-TR-73- (to be published).

LEWIS DISCUSSION

LAPP - I am quite impressed that you can get depolarization data
for this type of experimental situation. The data seems to me to be of
high value, because it tells you something about the clustering process
during the course of a fluid mechanic expansion.

BARRETT - Is that the first time it has been reported?

LEWIS - I think so.

LAPP - I believe that evaluations of Raman scattering diagnostics
to fluid mechanic problems such as the one discussed by Lewis are very
rewarding. Both Lewis and Ron Hill at Sandia have been involved in
programs to compare Raman scattering and electron beam-induced fluorescence
in a density region of overlapping applicability, and the results so far,
I hear, are encouraging.

STUDY OF REACTIONS PRODUCING VIBRATIONALLY-EXCITED NITROGEN*

by

Graham Black

Stanford Research Institute
Menlo Park, California

ABSTRACT

The use of Raman scattering to measure the production of vibrationally-excited, infrared-inactive diatomic molecules in chemical reactions is described. Examples based upon the observation of nitrogen anti-Stokes scattering are given.

* * * * * * * * * * * * * *

We use Raman spectroscopy to measure the production of vibrationally-excited but infrared-inactive diatomic molecules in chemical reactions. To date our measurements have been confined to vibrationally-excited nitrogen produced in the reactions

$$O(^1D) + N_2 \rightarrow O(^3P) + N_2 + 1.96 \text{ eV} \tag{1}$$

$$N(^4S) + NO \rightarrow N_2 + O(^3P) + 3.27 \text{ eV}. \tag{2}$$

The aim of the experiments is to determine the fraction of the available energy which appears in nitrogen vibrations. It is expected that the initial vibrational distribution will be rapidly degraded by vibrational exchange collisions of the type

$$N_2(v=n) + N_2(v=0) \rightarrow N_2(v=n') + N_2(v=n-n'), \tag{3}$$

*Work sponsored by the Advanced Research Projects Agency through Contract DAHC04-72-C-0015 with the U. S. Army Research Office, Durham.

resulting in an increased population of $N_2(v=1)$. Hence a measurement of $N_2(v=1)$ should be sufficient to determine the vibrational energy in the nitrogen.

The measurements have been made by observing Raman scattering of 4880 Å Ar^+ laser radiation. The intensity of the Q-branch of the anti-Stokes transition (at 4382 Å) was measured, because the anti-Stokes spectra are free from Raman transitions from the $N_2(v=0)$ level [whereas the Stokes transitions from $N_2(v=1)$ lie in the rotational structure of the Stokes lines from $N_2(v=0)$]. With a 1 Å bandpass on a Spex 1402 double monochromator, using 700mW of 4880 Å radiation and collecting the Raman light over 1 steradian, an experimental sensitivity in excess of 100 counts sec^{-1} [torr $N_2(v=1)]^{-1}$ has been obtained using an EMI 6256S photomultiplier. This calibration is obtained by observing the signal from heated nitrogen.

Photon counting with long integration times and subtraction of back-ground signals have been used to determine signals which have been in the 10^{-2}-10^{-1} counts sec^{-1} range when studying reaction (1) and in the 1-10 counts sec^{-1} range when studying reaction (2) (in the presence of a dark count of 5 counts sec^{-1}). The major reason for the difference in signal levels arises from the much smaller amounts of $O(^1D)$ (made by the photolysis of O_2 at 1470 Å) compared to $N(^4S)$ (made by a microwave discharge in N_2). The lowest signals measured corresponded to $N_2(v=1)$ pressures $< 10^{-3}$torr. See Reference 1 for a complete description of this work.

Our present experimental program is also concerned with the wall deactivation of $N_2(v=1)$ on a variety of different solid surfaces.** Again, the intensity of the Q-branch of the anti-Stokes Raman scattering at 4382 Å (using 4880 Å laser radiation) is used to monitor the $N_2(v=1)$ concentration. The vibrationally-excited nitrogen is produced by a thermal source and by a microwave discharge. The results are interpreted in terms of the two-dimensional diffusion equation with Poiseuille flow. The two sources of $N_2(v=1)$ give somewhat different values for the wall deactivation coefficient γ. Furthermore, the results with the microwave source depend on the length of exposure of the surface to the afterglow. The observed differences are probably related to the fact that the microwave source also produces atoms and the thermal source does not. The lowest values of γ have been recorded for quartz and Pyrex after 24 hours of exposure to the afterglow ($1.8x10^{-4}$ and $2.3x10^{-4}$ respectively) and the highest for an aluminum alloy (type 5052) under the same conditions ($5x10^{-3}$). The results are being interpreted in terms of a mechanism of heterogeneous vibrational deexcitation.

---

1. G. Black, R. L. Sharpless, and T. G. Slanger, J. Chem. Phys. <u>58</u>, 4792 (1973).

**Work supported by Contract F33615-72-C-1613 with Wright Patterson AFB.

APPLICATION OF RAMAN SCATTERING TO
GASDYNAMIC FLOWS*

by

J. W. Daiber

Calspan Corporation

ABSTRACT

We discuss application of Raman scattering diagnostics to several
practical measurement situations. These situations involve turbulent
mixing in high-pressure nonsteady flows, short-duration steady flows,
chemical and electrical excitation of laser media, and mixing in
atmospheric simulation studies. The capabilities of Raman scattering
techniques are compared to those of alternative techniques, and some
experimental results are presented.

I. INTRODUCTION

The primary attraction of Raman scattering as a diagnostic appears
to be its ability to provide localized measurements of species concentrations and the distribution of these species in their internal energy
states. At Calspan we are involved with many types of nonsteady and
short-duration flows where such information is needed and appears to
be obtainable from a Raman scattering probe. Representative applications are presented in the following sections. First, an application
is described in which Raman scattering was used for studying nonsteady
flows. Second, the constraints placed on a Raman diagnostic probe by
the short duration and low densities found in aerodynamic testing
facilities are discussed. Third, a description is given of an
infrared emission probe used for species determination in steady flow
lasers. Fourth, the possibility of using Raman scattering to follow
mixing processes in the Calspan Atmospheric Simulation Facility is
suggested.

---

*This paper was written with support of the Office of Naval Research
under Project SQUID.

II. TURBULENT MIXING STUDIES[**]

A study has been made at Calspan of the turbulent mixing of hydrogen and helium at high pressures.[1] The anticipated pressure levels were as high as 5000 psi which made Raman scattering a particularly attractive diagnostic for following the time history of the mixing. A similar study of the mixing of nitrogen and helium was subsequently carried out by Hartley.[2]

The hydrogen was introduced into the helium chamber via a jet formed by 3/16-inch diameter tubing. An AVCO pulsed nitrogen laser was focused into the chamber. The light scattered at 120° was detected by an RCA 8575 photomultiplier tube with an interference filter centered to pass the complete hydrogen Raman band at 392.1 nm. The number of photons scattered into the detector can be calculated from the formula

$$N_s = N_i \, n \, L \, \sigma \, \Omega \, T$$

where N_i is the number of photons in the incident radiation, n the number density of the scattering molecules, L the length of incident beam viewed by the detector, σ the scattering cross section of the molecule, Ω the solid angle collected, and T the transmission factor that includes all losses due to absorption and reflection in the optical path.

Assuming a laser output of 200 kW, one obtains for a pulse duration of 10 nsec and a wavelength of 337.1 nm an incident number of photons equal to 3.39×10^{15} photons/pulse. The number density of scattering molecules at 1000 psi and 300°K is 1.67×10^{21} cm^{-3}. The scattering cross-section was estimated[3] as 7×10^{-30} cm^2/sr. The collection angle was calculated to be 0.29 sr and the length viewed by the detector was 0.2 cm. This corresponds to the viewing region being imaged onto a 0.635-inch diameter aperture by a 25-mm diameter lens of 19.5-mm focal length. The transmission factor was computed as the product of a transmission factor for the 392.1-nm filter of 0.23, an attenuation factor of 0.90 for the aspherical laser lens (estimated), and a total reflection loss factor at eight surfaces (two sapphire windows and two lenses) of $(0.96)^8 = 0.719$, giving an overall factor of 0.149. The number of photons collected per laser pulse is then 388,000. Assuming a photomultiplier quantum efficiency of 28%, the number of photoelectrons created per laser pulse is 109,000. The minimum statistical noise in a signal is given by

[**]This research was supported by the U.S. Atomic Energy Commission under subcontract from the Sandia Corporation.

$\sqrt{N_s}/N_s$, which for this case amounts to 0.3%. Based on this analysis one would expect the signal to be much greater than the noise and, therefore, that the mixing could be followed.

A typical set of data is shown in Figure 1. The initial mixing is completed in approximately 0.2 sec. This is followed by a slower mixing during which all the pockets of hydrogen are dissipated. It was difficult to follow the turbulent mixing because the noise in the Raman signal was greater than that estimated above. This result indicated that other sources of noise are dominant.

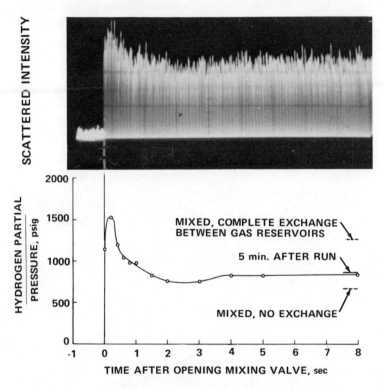

Figure 1.   Mixing of hydrogen (4080 psig) into helium (1940 psig).
The Raman probe is focused on the centerline of the
gas jet. The upper figure is the Raman scattered
signal as recorded on the oscilloscope. The lower
figure is the corresponding hydrogen pressure variation.

The incident laser pulse was simultaneously monitored; however, its pulse-to-pulse variation while not negligible did not appear to correlate with those in the scattered light. Rayleigh scattered light in the wings of the interference filter is a possible source of signal due to random appearance of particulates in the scattering volume. This possibility was monitored by making simultaneous measurements using a second filter that was centered at 365.8 nm; no significant signal

was observed. A further test was made to ascertain whether the fluctuations were due to random variations in the plane of polarization; as
expected for this laser they were not.

The recording system effectively integrates the 10 nsec duration
laser pulse. Therefore, the number of photoelectrons per pulse is
given by $N = UC/Ae$, where U is the peak amplitude of the pulse (volts),
C the capacitance of the photomultiplier circuit (575×10^{-12} farads),
A the current amplification of the photomultiplier (4×10^6), and
e the electron charge (1.6×10^{-19} coulombs). The minimum deviation
in signal is then $1/(30\sqrt{U})$. Experimental results obtained in **several**
static calibration runs were evaluated in terms of mean values $\bar{x}$ and
standard deviations σ_D, taking in each case twenty readings of
individual pulse heights. The relative quantities $\sigma_D/\bar{x}$ were then
plotted against mean pulse height U, and compared with the theoretical
values. A plot obtained from several calibration runs performed in
hydrogen is shown in Figure 2. The standard deviations are seen to
exceed the theoretical minimum by roughly a factor of three. The
tendency to approach the theoretical value for the largest signals is
caused by saturation.

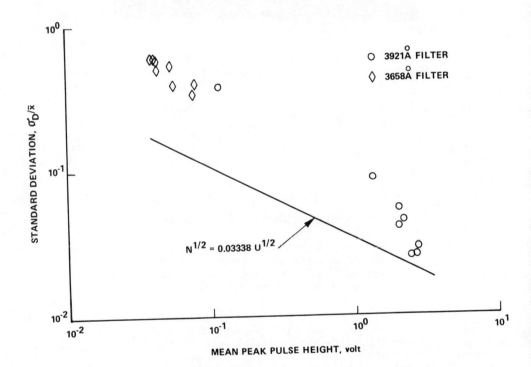

Figure 2.   Standard deviation of the peak pulse height for
             Raman scattering in hydrogen. Laser operated at
             50 pulses/sec, 9.5 kV, and 20 Torr.

The data shown in Figure 1 have an average signal level of 1.1 volt. Using the calibration curves derivable from Figure 2, this corresponds to a pressure of 720 ± 100 psi. The final partial pressure of hydrogen measured at very late times was 860 psi. The approach to equilibrium cannot be followed with certainty because the variation in the signal corresponds to pressure fluctuations of 200 psi. Thus the Raman scattering was able to measure gross effects in the flow field.

Improvements in this system are possible. The collecting angle could be enlarged, a more powerful laser used, optics could be anti-reflection coated, a cooled or less noisy photomultiplier used and, most importantly, a more reproducible laser source employed. With these modifications and other improvements the high-pressure gas-mixing process has been followed using Raman scattering.[4]

III.   SHORT-DURATION STEADY FLOWS

In studies involving high-temperature reacting flows it is often essential that the concentrations of species be accurately known. This is always the case for those experiments designed to measure chemical or vibrational reaction rates. However, such experiments can be tailored to a particular diagnostic which will yield the required concentrations with the necessary accuracy. When tests are run in large aerodynamic facilities for the simulation of flow fields, then the measurement of species density can be very difficult. There are presently two programs in progress at Calspan which could have utilized a nondisturbing diagnostic for the remote measurement of species concentration. These are rocket exhaust plume studies in the high-altitude chamber and external burning studies in a hypersonic shock tunnel.

A schematic diagram of the rocket exhaust-plume in an altitude chamber is shown in Figure 3. The research engines are similar to actual flight engines and employ the same propellant combinations. In these experiments, an identification of species and a point measurement of species concentrations in the exhaust plume are needed for validation of computer predictions of the flow field and plume radiance. Important exhaust species for a typical solid propellant are listed in Table I. Various other species may be important for specific propellants, such as Cl_2, O_2, NH_3, NO, NO_2, and HF. Typical concentrations of the dominant species in the exit plane of the rocket are estimated to be the order of 10^{17} cm^{-3}. For other test conditions the exit densities could be a factor of ten smaller. The duration of these tests is approximately 20 msec after which wall effects can perturb the flow in the regions of interest.

Prediction of the radiation emitted by the plume requires that both the species concentration and excitation temperatures be known. The rotational temperature can be determined from Raman scattering,[5] and in many cases it is the same as the species translational

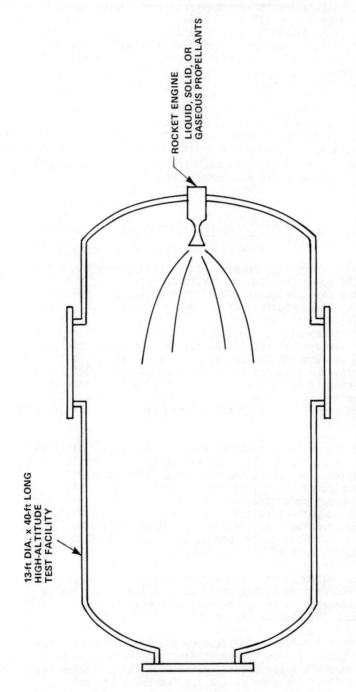

ROCKET ENGINE
LIQUID, SOLID, OR
GASEOUS PROPELLANTS

13-ft DIA. x 40-ft LONG
HIGH-ALTITUDE
TEST FACILITY

Figure 3.   Schematic diagram of rocket exhaust plumes being generated in the high-altitude
test chamber.

Table I

Testing Condition for Rocket Engine Firings
with Solid Propellant

CHAMBER PRESSURE	3000 psia
CHAMBER TEMPERATURE	3500°K
EXIT AREA RATIO	30
EXIT DENSITY	10^{18} cm^{-3}
EXIT SPECIES H_2	30%
CO	20%
HCl	15%
H_2O	15%
N_2	10%
CO_2	5%
PARTICULATES	5%

temperature. The distribution of species in the vibrational energy
states controls the infrared radiation. The distribution can be
found by analyzing the shape of the Stokes hot bands[6] or the vibrational
temperature can be determined from the Stokes to antistokes
intensity ratio.[7]

The external burning experiment is shown in Figure 4. A flat
plate model is placed in the shock tunnel so that it intercepts the
hypersonic flow at the desired angle of attack. At a longitudinal
station on the model a row of nozzles inject fuel into the hot air-
stream. In these experiments it is necessary to know the local fuel/oxi-
dizer ratio and the degree of completion of the combustion reactions
in order to develop and verify analytical models of the process. Full
interpretation of the data also requires that the combustion temperature be determined. Typical densities in the combustion region are
10^{17} cm^{-3} and the tunnel flow lasts for 2 to 5 msec.

At present, the species concentrations are determined by collecting
samples with quick closing probes and then analyzing the gases in a
mass spectrometer. This technique has several disadvantages. First,
the probes perturb the flow field. Second, hydrogen is used as the shock
tube driver and some of this gas could enter the collecting probe.
Third, the short flow time precludes traversing the probes during a
run, thus necessitating that several runs be made to obtain a spatial
distribution. Since shock tunnel experiments are very costly, an
alternate approach is highly desirable.

The potential of Raman scattering for obtaining the required data
for both of these test programs is obvious. The simultaneous
determination of local temperatures and concentrations at several
spatial locations would add substantially to the value of the experiments.
However, the difficulty in doing this becomes apparent when typical
signal levels are estimated.

Four types of laser sources are commonly employed in Raman studies.
The minimum detectable concentrations of CO_2 will be compared on the
basis of a 1 msec pulse from a 25 W argon laser, a 10 nsec pulse from a

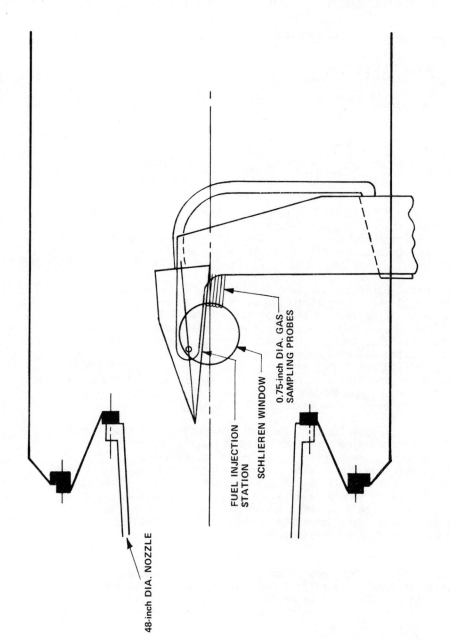

48-inch DIA. NOZZLE

FUEL INJECTION STATION

SCHLIEREN WINDOW

0.75-inch DIA. GAS SAMPLING PROBES

Figure 4.   Measurement of gas concentrations on the external burning model in the 96-inch diameter hypersonic shock tunnel.

500 kW nitrogen laser, a 10 joule pulse from a ruby laser, and a
frequency doubled ruby laser with a 10% efficient doubling crystal.
The minimum usable signal-to-noise ratio is taken to be 20. It is
assumed that the detector is not noise limiting so that only 400
photoelectrons need be observed. The transmission efficiency is taken
to be 25%. It is further assumed that the collecting mirror is one
foot in diameter and located six feet from the scattering center.
Finally, a spatial resolution of one centimeter is considered sufficient.
The smallest detectable CO_2 concentration is found by substituting these
variables into the equation given previously for the scattered signal.
The results of the calculation are summarized in Table II.

These calculations indicate that the largest signals should be
obtainable by using a doubled ruby laser source. The scattered intensity
should be sufficient to detect major species in the flow fields of the
external burning experiments. However, there will be many practical
problems in carrying out such a measurement. The combustion region
is luminescent and photons at the Raman Stokes and antistokes
frequencies will be generated. The intensity of this background
emission will be high and this radiation will need to be carefully
discriminated against. An additional problem will be in achieving a system so that 400 photoelectrons in the detector are indeed sufficient to
give interpretable signals that are adequately free of noise.

IV. STEADY STATE LASER STUDIES[***]

While the facilities discussed so far are of the short test
duration type, there are facilities at Calspan in which running times
are very long. These include the chemical and electrical lasers. In
these programs, the populations of the various vibrational energy
states must be measured for comparison with theoretical models. At the
present time, this is done by synchronous detection of infrared
emission using a scanning monochrometer.[8] The rotational temperature is
determined from spectra showing resolved rotational lines.[9] For
diatomic gases, the populations of the vibrational states are then
found by comparing computer generated spectra with experimental spectra
taken with wide slits so that individual lines are not resolved.[8]
A set of reduced data is shown in Figure 5. The population of the v = 14
state is approximately 10^{12} cm^{-3} and represents the lower limit of the
present equipment.

The ir technique integrates across the flow field; the advantage of
using a Raman scattering diagnostic is that point measurements of
species concentration can be made. In a chemical laser the spatial
resolution could give information on mixing phenomena. The second
advantage for any type laser is that ground state populations or non-
radiative intermediate species can also be monitored.

[***]This research was carried out under contract to the Air Force
Avionics Laboratory, AFSC.

Table II

Minimum Detectable CO_2 Concentrations

LASER	ARGON	NITROGEN	RUBY	DOUBLED RUBY
WAVELENGTH, nm	488.0	337.1	694.3	347.2
PERFORMANCE	25 W	500 kW	10 j	1 j
PULSE DURATION, sec	10^{-3}	10^{-8}	2×10^{-8}	2×10^{-8}
INCIDENT NUMBER OF PHOTONS	6.2×10^{17}	8.5×10^{16}	3.5×10^{20}	1.7×10^{19}
OPTICS EFFICIENCY	0.25	0.25	0.25	0.25
SCATTERING LENGTH, cm	1.0	1.0	1.0	1.0
RAMAN VIBRATION CROSS SECTION, cm^2/sr	8.1×10^{-31}	3.6×10^{-30}	2.0×10^{-31}	3.2×10^{-30}
COLLECTING ANGLE, sr	0.022	0.022	0.022	0.022
DETECTOR EFFICIENCY	0.21	0.32	0.055	0.34
PHOTOELECTRONS PER SCATTERING MOLECULE	5.7×10^{-16}	5.3×10^{-16}	8.3×10^{-15}	1.0×10^{-13}
MINIMUM CO_2 DENSITY, cm^{-3}	7.0×10^{17}	7.6×10^{17}	4.8×10^{16}	4.0×10^{15}

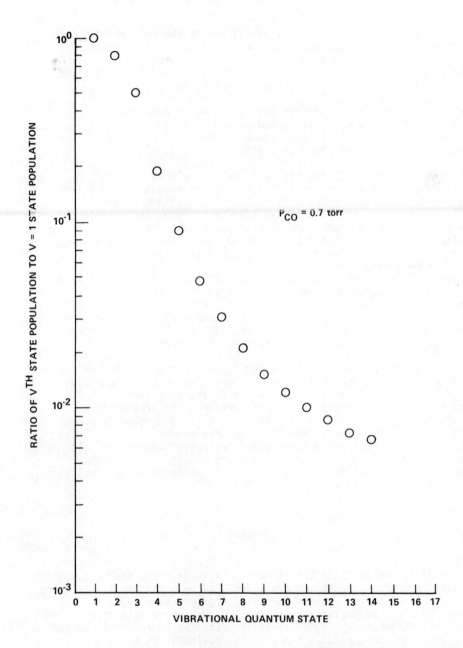

Figure 5.   Experimental vibrational population distribution data
for a cryogenically cooled CO electrical discharge
laser.

V. ATMOSPHERIC MIXING STUDIES

One of the newest wind tunnels at Calspan is the Atmospheric
Simulation Facility, Figure 6. This microsonic tunnel can generate
air speeds of up to 80 mph in a 6-ft high and 8-ft wide test section.
For the simulation of atmospheric phenomena, it is used at low speeds
with specially designed tunnel floors for generating the correct
scale of turbulence and spatial distribution of turbulence for the
particular problem. Test programs have so far been concerned with
building-induced pedestrian-level winds, wind loads on structures,
and effluent profiles from exhaust stacks. This latter program is one
in which Raman scattering could be useful.

To simulate correctly the mixing of exhaust plumes with the
atmosphere, the buoyancy and exit momentum of the exhaust gas must be
scaled. What this usually means is that, to work with models of
suitably small scale, the exhaust gas must be as light as possible.
At the present time, helium-hydrogen-nitrogen mixtures are used.
The helium is added in proportion to the SO_2 (or other contaminant)
concentration in the effluent. Downstream concentrations are
obtained by measuring helium concentration profiles with a mass
spectrometer which samples through a group of capillaries mounted on a
rake. With this procedure, the lower limit to the detectable helium
concentration is set either by the trace concentration in atmospheric
air, 4 ppm, or by the background rise during sampling. Thus the Raman
scattering probe cannot offer improved sensitivity, but it can offer
a nonperturbing probe for measurements near structures.

A more important measurement in the tunnel is the velocity of the
gas streams. This is presently done with hot wire probes. However,
these probes, unless temperature compensated, are sensitive to any
induced temperature fluctuations. In future programs, temperature
gradients will be simulated and therefore an alternate velocity probe
will be needed. There are presently under development at Calspan both
a double hot wire probe and a laser velocimeter for induced air velocity
studies. The extension of laser velocimetery combined with Raman
scattering into a probe for mixing studies in the Atmospheric Simulation
Facility appears attractive.

VI. SUMMARY

Four different applications of Raman scattering have been discussed.
The limitations of the technique are associated with the small scattering
cross section. Even for studies involving high-pressure gases the
technique required sophisticated components. Low-noise detectors
combined with high-energy, high-repetition rate and reproducible pulsed
lasers could, however, greatly improve the applicability of this
diagnostic. At Calspan, work is in progress to extend the Raman
scattering diagnostic to quasi-continuous testing facilities.

Figure 6. Model of power station with coal piles and cooling towers being tested in the atmospheric simulation facility.

REFERENCES

1.   G. H. Markstein and J. W. Daiber, "Turbulent Jet Mixing" Calspan
     Report No. CM-2637-A-1 February 1970.

2.   D. L. Hartley, "Transient Gas Concentration Measurements
     Utilizing Laser Raman Spectroscopy" AIAA J. 10, 687 (1972).

3.   D. G. Fouche and R. K. Chang, "Relative Raman Cross Sections for
     O_3, CH_4, C_3H_8, NO, N_2O, and H_2" Appl. Phys. Lett. 20, 256 (1972).

4.   D. L. Hartley, "Raman Gas Mixing Measurements and Ramanography,"
     this Proceedings.

5.   J. R. Smith, "Transient Flow-Field Temperature Profile Measurement
     Using Rotational Raman Spectroscopy," this Proceedings.

6.   M. Lapp, L. M. Goldman and C. M. Penney, "Raman Scattering from
     Flames," Science 175, 1112 (1972).

7.   S. Lederman, "Molecular Spectra and the Raman Effect, A Short
     Review," Pibal Report No. 71-15 (June 1971).

8.   J. W. Rich and H. M. Thompson, "Infrared Sidelight Studies in the
     High-Power Carbon Monoxide Laser," Appl. Phys. Lett. 19, 3
     (1971).

9.   G. Herzberg, Molecular Spectra and Molecular Structure, I. Spectra
     of Diatomic Molecules (D. van Nostrand Co. Inc., New York, 1950).

DAIBER DISCUSSION

WHITE - With respect to velocity, the obvious suggestion for a
measurement technique is laser velocimetry.

DAIBER - Yes, that is a very convenient diagnostic tool, but I
believe that you usually need to seed the flow to get enough scatterers.

WHITE - Depending on what the flow is, if high data rates are not
needed, it is a rare flow that does not have enough particles in it to
work with.

DAIBER - What we really need to know is how well the exhaust gas
mixes with the free stream air. One then has to worry about slippage
of the particles unless you have very small particles, and also in which
stream that particular particle is in. It would be nice if you could
simultaneously measure velocity and Raman scattering so that you knew if
you were looking at the stack gas or free stream air.

HENDRA - Is it possible to put a tracer material in the plume gas or to introduce an imitation plume gas?

DAIBER - Yes, you can mix in anything; however, the overall gas buoyancy and momentum has to be correct for the simulation.

REMOTE RAMAN TEMPERATURE MEASUREMENT
WITH A DYE LASER

by

M. E. Mack

United Aircraft Research Laboratories*
East Hartford, Connecticut

ABSTRACT

The use of dye laser excitation sources for Raman scattering temperature measurements is discussed. Experimental data for nitrogen in
air is given which shows good agreement with theory.

* * * * * * * * * * * * * *

Generally, remote scattering measurements have been carried out using
either the gaseous argon or nitrogen lasers or the solid state ruby or
neodymium lasers. Recent developments now make the dye laser an attractive alternative. Most attractive for remote sensing applications are the
repetitively pulsed flashlamp pumped dye lasers. The pulse durations are
typically of the order of a microsecond and pulse energies of one joule
or thereabouts are also typical. Repetitively pulsed average power outputs up to 10 watts have been achieved.[1] Overall electrical efficiency
is quite high relative to the solid state and gaseous lasers. Tuned
output efficiencies of 0.1 to 0.3% can be readily achieved.

The dye laser also offers an advantage in its tunability. This can
be an important advantage in attempting to avoid interfering fluorescences.
By the proper choice of dye, laser output can be achieved anywhere in the
visible spectrum, although with the dyes currently available high power
(one joule) output is achievable only in the 4500 to 4800 Å and 5800 to
6400 Å spectral ranges. A given dye can be tuned over perhaps 400 to 800
Å. Tuning is accomplished by using frequency selective feedback (grating,

*Present Address: Avco Everett Research Laboratory, Inc., Everett,
 Massachusetts 02149.

Fabry-Perot etalon, etc.) in the laser cavity. The linewidth will also depend on the degree of dispersion in the cavity. Tuned linewidths of 1 to 5 Å are readily achievable. Narrower linewidths can be achieved at some cost in power.

The dye laser does have two possible disadvantages for remote sensing applications. The first is that in contrast to the gas lasers, for example, the fluorescence background from the laser is continuous rather than discrete. The background can be blocked with a suitable filter but the frequency spread of the fluorescence is then the linewidth of the filter. This is not a problem for large shift scattering but would be for applications where the frequency shift is, say, 50 cm^{-1} or less. The second drawback of the dye laser is that, at least with flashlamp pumping, the pulse duration is too long for accurate range resolution. However, for remote vibrational Raman scattering where range resolution is not critical, the dye laser is ideal.

In the present case the dye laser has been used for laboratory measurements of gas temperature and species concentration. The experimental apparatus is shown in Figure 1. The laser system used is described elsewhere.[1] For the present experiments the laser was tuned to 5900 Å, produced a linewidth of about 5 Å, and achieved an output of about 0.1 joule per pulse. The gas sample was nitrogen in air obtained from a laboratory air line. For the temperature measurements the air passed through an electric gas torch, while for the species concentration measurements the air was admixed with argon and blown into the sample region. The actual sample region was about 0.04" by 0.04" transverse to the laser beam by about 0.15" along the laser beam. An RCA 8575 photomultiplier was used for the anti-Stokes (temperature) measurements while an RCA C31034 was used for the Stokes (species concentration) measurement. The interference filter bandpasses were about 30 Å.

Figure 2 shows the results of anti-Stokes measurement. The experimental data give an excellent fit to the theoretical curve. The 1/T factor in the theoretical expression takes account of the decrease in density with increasing temperature.

At room temperature a spurious signal was observed with an amplitude of about 0.02 units (see Figure 2). Eventually this was traced to the presence of particulates in the gas stream. Inserting a 10μ pore size filter in the gas stream nearly eliminates the signal. Blue cutoff filters in the laser beam do not affect the signal. Room air also produces the same spurious signal. It is interesting to note that this signal occurs on the anti-Stokes side of the exciting line. As would be expected, a much stronger interfering signal is also found on the Stokes side. Both signals are broad band as determined by tuning the dye laser across its 400 Å tuning band and observing the signal with fixed frequency interference filters.

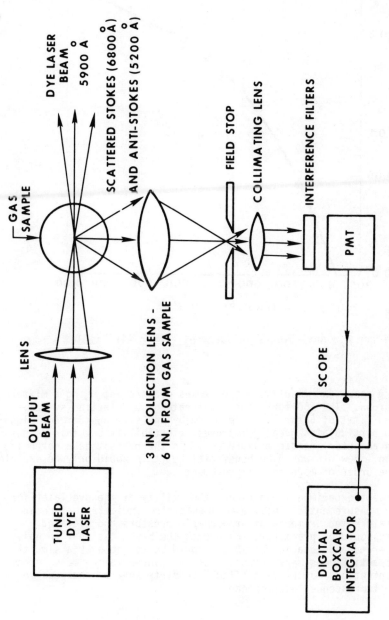

Figure 1. Experimental Apparatus

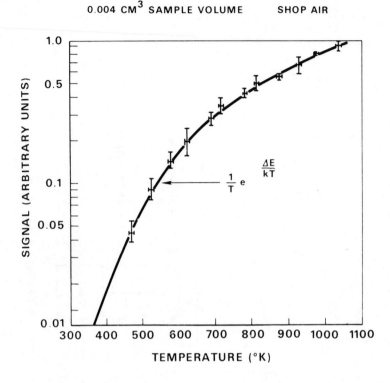

Figure 2.   Anti-Stokes Measurement - Shop Air Sample

Figure 3 shows the results of the Stokes measurement using a mixture of air and argon. The 150 torr signal saturates the detection system. The detection limit of 5 torr is determined by the spurious aerosol signal. Removing the particulates greatly improves detectibility but this is not practical in an actual remote measurement. In the present case narrowing the detection linewidth would be beneficial but this would only be expected to give a one order of magnitude improvement.

The present experiments underscore the utility of the dye laser for remote Raman measurements. These experiments also indicate that remote Raman temperature measurements at atmospheric pressure and elevated temperatures should be practical. Even with the present signal to noise ratios, temperature accuracies of ±0.3% should be possible by a simultaneous Stokes/anti-Stokes measurement. Species concentration measurements at low concentrations may prove difficult in dirty environments due to particulate fluorescence interference.

760 TORR TOTAL PRESSURE

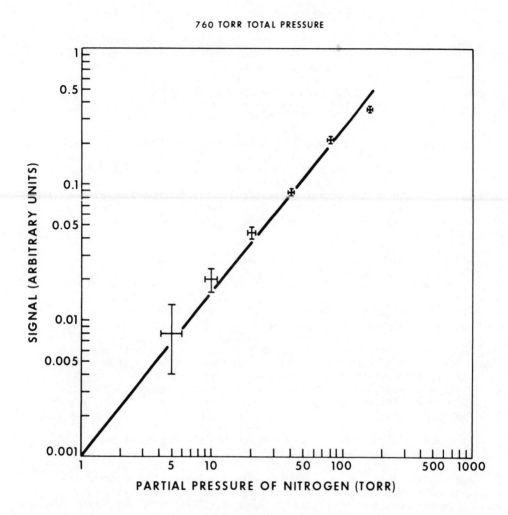

Figure 3.   Stokes Measurement - Shop Air and Argon Sample

1. M. E. Mack, Appl. Optics, January 1974 (to be published).

MACK DISCUSSION

MOREY - I wonder if you could describe your dye laser system?

MACK - We hope to get 1 joule pulses at 50 pulses per second. It is a proprietory sort of gadget, and is not commercially available. I think you are going to find a lot of companies coming out with high power dye lasers in a year or so. This unit has a very fast gas flow through the flash lamp which runs above atmospheric pressure. The lamp has a very long lifetime and will run quite readily at high repetition rates.

BERSHADER - What is the spectral width for this source?

MACK - The spectral width for these measurements was 5 Å which isn't particularly narrow. On the other hand, if you are measuring temperature by use of the Stokes/anti-Stokes ratio for nitrogen over a thousand or so degrees then the shift in the line center is of the order of, say, 5 Å, so you can't use interference filters that are any narrower than that anyway.

WHITE - There has been discussion comparing the merits of inter-ference filters and spectrometers. One of the disadvantages of inter-ference filters, of course, is their inability to be tuned. And this points out perhaps one of the advantages of dye lasers. If you want to go on and off resonance with a single interference filter, it can be accomplished simply by tuning the laser.

MACK - Yes, that's correct, and in fact the way we identified anti-Stokes scattering to make sure we had what we wanted, was that we simply tuned the laser away so the anti-Stokes radiation didn't pass through the filter anymore. That is a very nice advantage of tunable lasers, and again, when flourescence is a problem, it is possible to tune between the lines.

LEONARD - Did you have to make any special efforts to filter out the flourescence in the laser itself?

MACK - Yes, this can be a problem with dye lasers. We didn't par-ticularly have this problem because the dye laser beam was being used about 5 meters away from the actual laser. But it is true, as I men-tioned in my talk, that dye lasers have a continuous flourescent back-ground around the laser line. Although this background is down perhaps six orders of magnitude, filtering is still sometimes necessary.

LEONARD - I think that perhaps the dye could be the source of your "anti-Stokes" flourescence.

MACK - Well, that could have been a possibility. However, we put a valve into the air line and had one line with a filter in it and one without it, and you could switch between them and watch the anti-Stokes signal go up and down.

LEONARD - If the flourescence does originate from the laser, would it not also be scattered in proportion to the number of particles?

MACK - Yes, but we also put Corning sharp cut-off filters in the beam on the anti-Stokes side. Of course, you have to be careful with those too, because they fluoresce as well.

RAMAN SPECTROSCOPY OF SOME HYDROCARBONS AND FREONS
IN THE GAS PHASE

by

David A. Stephenson

Research Laboratories, General Motors Corporation

ABSTRACT

The usefulness of Raman Scattering for monitoring sources of simple gases such as CO, NO, SO_2, CO_2 has been demonstrated. Some hydrocarbons such as methane and benzene also give Raman spectra which are simple and separated well enough from other species to make analysis straightforward. However, other hydrocarbons, no less important, such as toluene and pentane, have more complicated spectra. Spectra for various hydrocarbons are presented with scattering cross sections normalized to N_2. Spectra of various freons are also included and their use as internal standards for mass measurements discussed.

* * * * * * * * * *

The potential usefulness of Raman scattering to the measurement of combustion gas densities at low concentration has been demonstrated by the field work of Inaba and Kobaysi[1] in Japan and the laboratory work of Leonard[2] and Fouche, et al,[3] as well as others in this country.

The advantages of Raman scattering are fairly well known. They are:
    Range information,
    Single ended devices possible,
    Relatively low interference (where we mean, in particular,
        that the interference from water vapor is low).

Some of the disadvantages are:
    Weak signal,
    Good scattered light rejection needed,
    .. at least 10^6 for N_2,
    .. at least 10^{10} for elastic or Tyndall scattering.

One of the least appreciated disadvantages is that fairly high resolution is often needed. This conflicts with the requirements of high sensitivity and good rejection of unwanted frequencies.

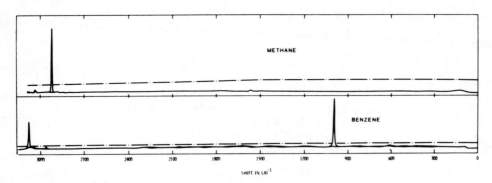

Figure 1.

    Here I would like to discuss the spectra of some hydrocarbons and
the problems which arise when we try to detect them at low concentrations.
All the data shown were taken with 1 watt of 4880 Å radiation. The spectra were recorded with a Spex Ramalog with 1800 lines/mm gratings.

    Figure 1 shows spectra of methane and benzene. The dashed curve
shows the instrument response for an equal concentration of N_2. All the
data shown are for one atmosphere or the vapor pressure at room temperature, whichever is smaller, along with the N_2 instrument response. Cross
sections for the various peaks can be obtained relative to the absolute
N_2 cross section from the response curve. The spectra here are fairly
simple and one could reasonably hope to build a system to detect them.
We see, however, that the C-H stretching region around 3000 cm^{-1} is not

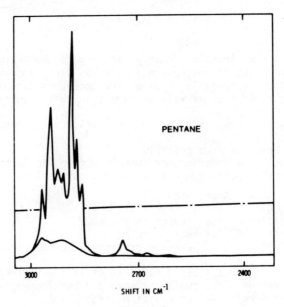

Figure 2.

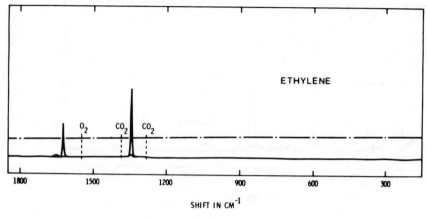

Figure 3.

going to be very useful because all hydrocarbons have lines here and some
of them, such as butane and pentane, can be quite complex and cover the
whole range occupied by the other, as shown in Figure 2. We will thus
restrict our attention to the region below 1800 cm^{-1}.

Figure 3 shows the spectrum of ethylene. The dotted lines indicate
the position of the Q branches of O_2 and CO_2. For analyzing any combustion products (where CO_2 may be as high as 15%) the line at ~1342 cm^{-1},
less than 10 cm^{-1} away from the strongest CO_2 line, must be discarded.
This leaves the line at 1623 cm^{-1} which is right in the middle of the
vibration rotation band of O_2, and which therefore could present a
problem at low relative ethylene/oxygen concentration. Since the ethylene line is highly polarized and the vibration rotation bands of
oxygen are strongly depolarized, it may be possible to get added discrimination against oxygen by rotating a polarization analyzer in front
of the detector with synchronous detection.

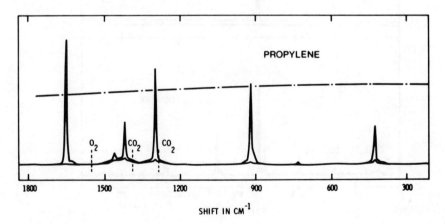

Figure 4.

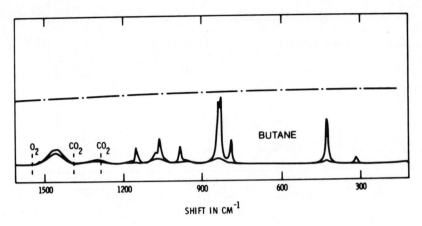

SHIFT IN CM^{-1}

Figure 5.

The spectrum of propylene is shown on Figure 4. It too has a strong line at 1620 cm^{-1} and thus could be detected along with ethylene, provided the O_2 interference is not too high. The spectra of the butylenes are similar and also have strong lines at 1620 cm^{-1}. It might seem that some of the lines at lower frequencies could be used for identification, but a look at the spectrum of butane in Figure 5 shows the difficulty with this. Butane and the other paraffins have very rich spectra in this region and will make the problem of trying to detect other hydrocarbons in this region very difficult.

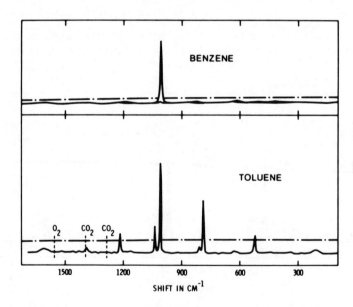

SHIFT IN CM^{-1}

Figure 6.

Figure 6 shows spectra of benzene and toluene. The line at 990 cm^{-1} which is common to both is fairly clear and may be used to identify them. To identify toluene separately is more difficult because butane again interferes with the line at 770 cm^{-1}.

In measuring sources of pollution, it is often desirable to measure mass flow and not just concentrations. One way to do this is to inject some internal standard such as freon at a known flow rate. The mass flow rate of the pollutant in question can then be obtained from the ratio of the concentration of the pollutant to that of the standard. Ideally, the standard should be inert, not normally present and easy to detect. Noble gases meet the first two requirements but not the third, at least for optical detection. Freons appear to be likely candidates for meeting these three criteria. The biggest question is that of stability or inertness in say a hot exhaust.

The spectra of some of the more promising freons are shown in Figure 7. The spectra of most are more complicated than one would like, thus causing possible interference with hydrocarbons. Freon 14 or tetra-

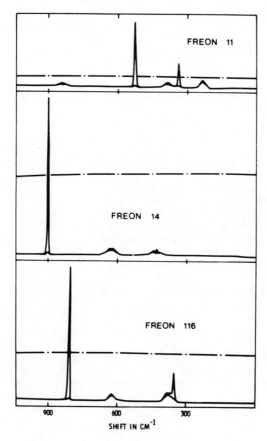

Figure 7.

fluromethane looks most promising from a spectroscopic point of view.
It is also one of the most stable. Not surprisingly perhaps, it is one
of the most expensive.

In conclusion, although the hydrocarbons have relatively large cross-
sections (on the order of 10 times that of nitrogen), the mutual interferences among them can be severe.

REFERENCES

1. T. Kobayasi and H. Inaba, "Spectroscopic Detection of SO_2 and CO_2
    Molecules in Polluted Atmosphere by Laser - Raman Radar Technique,"
    App. Phys. Lett. 17, 139 (1970).

2. D. A. Leonard, "Measurement of NO and SO_2 Raman Scattering Cross
    Sections," J. App. Phys. 41, 4238 (1970).

3. D. G. Fouche and R. K. Chang, "Relative Raman Cross Sections for N_2
    O_2, CO, CO_2, SO_2 and H_2S," App. Phys. Lett. 18, 579 (1971).

STEPHENSON DISCUSSION

LAPP - Was the data that you have shown as taken with your Spex
double monochromator?

STEPHENSON - Yes, and we have taken much more.

PENNEY - What are your conclusions regarding the hydrocarbons?

STEPHENSON - I don't think you can sort them out. Since last fall
I have looked at the total areas in a CH-stretch region. And if you
just want to look at total hydrocarbons the integrated cross sections relative to nitrogen, with the exception of acetylene, are all in the same
ballpark. But they go in the wrong direction for our purposes. Ethylene
is one of the weakest scatters, and you would really like to weight that
much more heavily. But the situation is not impossible.

PENNEY - How do you correlate to the numbers of CH bonds?

STEPHENSON - Fairly well for the heavier hydrocarbons, but acetylene
and ethylene produce significantly lower Raman scattering per CH bond.

HENDRA - There is a precise correlation between Raman intensity and
quantity of material, but of course you have to know the exact form of
the vibrational coordinates if at all possible and hence the value of
$\partial\alpha/\partial q$ for each mode of each constituent. People claim they can predict
these from first principles but of course they always observe them first!

Thus, I don't think it is possible to look at a spectrum of mixed alkanes
and say that the area under the Raman peak can be related to the amount
of total hydrocarbons. It cannot be done that way. However, an estimate
of this type is much closer to the right answer than one derived from
infrared absorption.

BRESOWAR - I would like to point out that total hydrocarbon analysis
is the chief technique used today in the exhaust emission work of aircraft
engine industries. A question that I would like to see answered is whether
Raman scattering can help with overall hydrocarbon measurements. It might
be satisfactory simply to correlate the C-H Raman band intensity to the
measurements of a conventional total hydrocarbone analyzer. I realize
that the C-H band intensity is weighted differently for different molecules
but the responses of present hydrocarbon analyzers are also non-uniform.

HENDRA - As long as we don't assume that it's going to be a very
quantitative tool as an indicator --

BRESOWAR - But it's quite possible that the weighting given by Raman
scattering is closer to what we really want to know. It is important to
realize that the measurement techniques we use now are far from perfect.
Anyone who believes in the precision of present measurements should have
no trouble accepting Raman scattering.

CHAMPAGNE - Pat Hendra is certainly correct in indicating that laser
Raman spectroscopy does not now provide a basis for making very quantitative, which is to say accurate, gas property measurements in situations
that are of engineering importance. But as Lt. Bresowar has noted, many
standard engineering instruments produce results that are far from perfect. For example, the literature of gas chromotography provides ample
evidence that the so-called total hydrocarbon analyzers, based on the flame
ionization detector, respond inconsistently to a large number of species
containing C-H bonds, including those of families as far removed from the
hydrocarbons as the alcohols and aldehydes. These devices can provide
very repeatable results in the sense of giving very much the same output
when presented with a given mixture. They certainly cannot be called
accurate since there is no definition of the extent to which the instrument is sensitive to each specie in a mixture. The accuracy of present
concentration measurements is often debatable even when one is using a
measurement mode which is known to be specific to the substance being
measured (such as using a good chemiluminescent analyzer for measuring
the concentration of NO in combustion effluent). The problems here are
those of representative sampling and sample transport from the parent
mixture to the analyzer. Present devices will not fulfill future needs
simply because they are neither specific nor accurate, or because extreme
care and great labor are required in sampling, calibration, and maintenance
of the devices to obtain accurate results. There should be no question
that a specie probe based on laser Raman spectroscopy will be a failure
if it is ultimately capable of no more than continuing the deficiencies
of present devices. Yet, I see no reason to be pessimistic at this stage
of development.

THE POSSIBLE APPLICATION OF RAMAN SCATTERING MEASUREMENTS
TO TURBULENT MIXING LAYERS

by

E. Storm

California Institute of Technology

ABSTRACT

Many flow configurations of engineering significance involve free
turbulent mixing. In this paper we discuss some of the characteristics
of this phenomenon, and the possibility of using Raman scattering to
examine its detailed behavior.

* * * * * * * * * * *

Since a large number of flow configurations of engineering significance are related to processes involving free turbulent mixing,
increased attention has been given to this problem over the past few
years. A typical case is a fully separated flow, where an important
element is the shear layer which develops behind the separation point.
Turbulent mixing with large density non-uniformities plays a very
important role in combustion, chemical mixing of different species, and
more recently in chemical lasers.

During this same period, there has been a continuing effort at the
California Institute of Technology directed towards the understanding of
heterogeneous turbulent mixing. The construction of a new experimental
facility and the subsequent investigations by Brown and Roshko[1] resulted
in significant contributions toward the unraveling of the important
mechanisms of such flows. A major breakthrough in this respect was the
documentation of the existence of large scale, orderly structures in an
otherwise fully turbulent mixing layer, and the realization that this
structure is an essential feature of the plane turbulent shear layer,
being responsible for most of the turbulent transport of mass and
momentum. While Brown and Roshko concentrated on the fundamental
aspects of equilibrium turbulent mixing between two streams of different
species (He and N_2) in a zero pressure gradient, the later work of

Rebollo[2] included investigations in an adverse pressure gradient.

Figure 1 shows the test section where the experiments of Ref. 1 and 2 were carried out. Helium and nitrogen streams are introduced into the

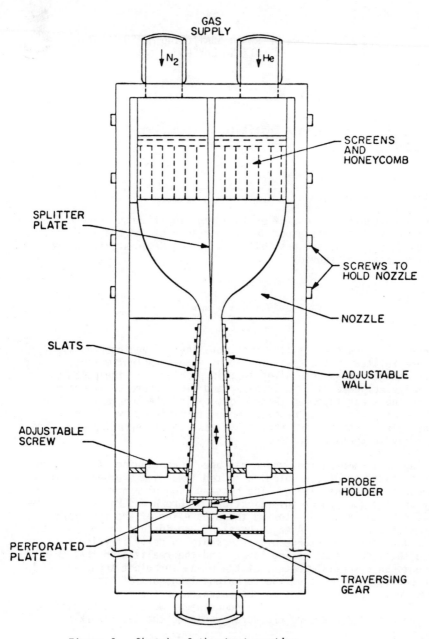

Figure 1. Sketch of the test section.

test section downstream of the splitter plate, most commonly subject
to the condition $\rho_{He}U_{He}^2 = \rho_{N_2}U_{N_2}^2$, the only case involving a non-zero
pressure gradient for which a similarity solution is possible. Thorough
measurements of both mean and fluctuating flow variables (static pressures, dynamic pressures and densities) have been carried out[1,2], and
the mean flow variables agree well, both quantitatively and qualita-
tively, with theoretical calculations based on the eddy diffusivity
and eddy viscosity model. Although the measurements of mean quantities
are continuing, and will be extended to different flow situations and
different density ratios, the main interest is currently directed towards
the fluctuating flow variables, and the density variations in particular.
The motivation for this is spectacularly demonstrated by Figures 2 and
3, showing the mixing layer model for the averaged values, and a shadowgraph of the instantaneous mixing layer respectively. The very large-
scale orderly structures are clearly visible, and correlation measurements show that even for fully turbulent flows, these "eddies" are
basically two dimensional.

Let us now focus our attention on the large scale shear layer
structure, and estimate the time scales involved in measuring density

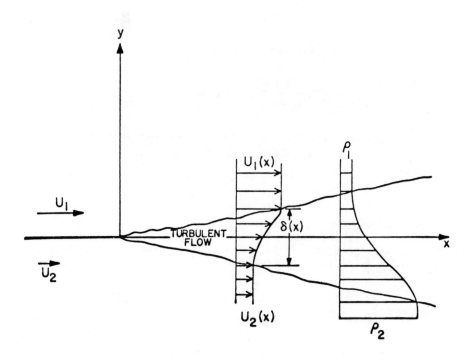

Figure 2. Mixing layer model.

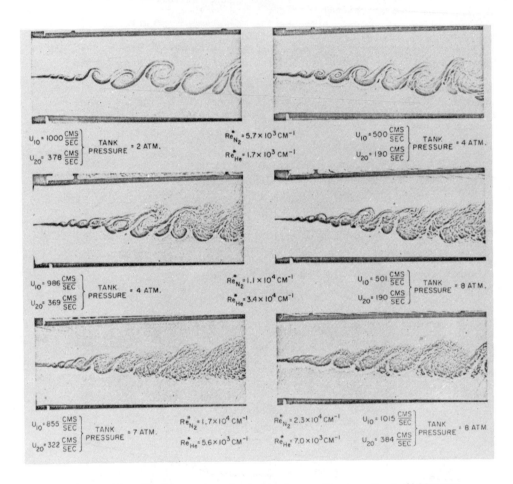

Figure 3A. Shadowgraphs of mixing layer for $\alpha = 0$ at different
Reynolds numbers. ($\rho_1 U_1^2 = \rho_2 U_2^2$ case.) Lower (low
speed) stream is N_2; upper (high speed) stream is He.
(Conditions correspond to zero pressure gradient.)

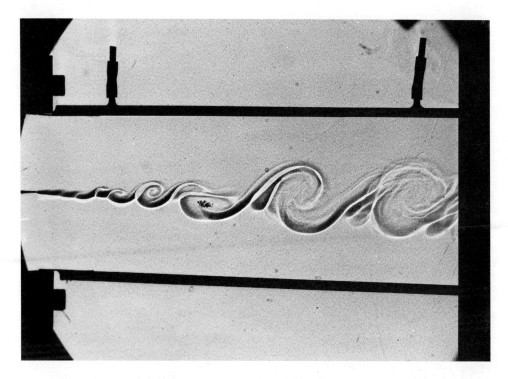

Figure 3B.   Detailed shadowgraph for same conditions as in upper
             left hand part of Figure 3A.

gradients and fluctuations through this layer. Using the width of the
trailing edge of the splitter plate ($\sim$1mm) as a typical value for the
smallest dimension associated with a "sheet" (see Figure 3) within the
shear layer, and noting the free stream velocity ($\sim$ 1000 cm/sec), we
arrive at an estimate of 10 kHz as a desirable value for the sampling
rate. Combining this with the desirability of a non-interfering
measuring technique strongly suggests Raman scattering as a diagnostic
tool.

     In principle, utilizing the Raman effect as a density probe is a
straightforward optical measurement. It simply involves focusing a
monochromatic light source of wavelength λ_0 on a point in the flow,
and measuring the amount of scattered light at a wavelength $\lambda_0 + \Delta\lambda$, where
$\Delta\lambda$ is uniquely determined by the scattering species. By the very nature
of the Raman effect, the intensity of the scattered light is directly
proportional to the density of scatterers, i.e. molecules in the focal

volume. The frequency response of the photomultipliers and other
detection equipment can easily be kept in the megacycle range, and the
focal dimension is, in principle, on the order of a few times   the wavelength of the incident laser light. All in all, it seems the perfect
method for a non-disturbing, instantaneous point measurement of
density. However, the invariance of difficulties in an experimental
measurement is still there, as anyone familiar with Raman spectroscopy
will be aware of. The Raman effect is a very weak one, with scattering
cross sections of the order of $10^{-30} cm^2$. The consequences of this
"weakness" will be apparent if we look at a typical situation in the
shear layer experiment.

Focusing an argon ion laser to a 0.1 mm spot in the shear layer,
and looking at the scattered light from nitrogen at 5 atmospheres, it
can readily be shown that a typical number of photons that can be
expected to be detected per second, per watt of incident laser power, is
of the order of 5×10^4. As we computed earlier, 10 kHz was a lower limit
on the necessary sampling rate. If we wish to say something definite
about the composition of one of these smallest "eddies", a sampling time
of 10-20 μsec would be desirable. Or in other words, we would like to
make a definitive measurement of the density of a point, averaged over
no more than 20 μsec. This process should be repeated after another
20 μsec, and so on. However, multiplying our two numbers of 20 μsec and
5×10^4 photons/sec/watt, gives a total of 1 detected photon per watt of
incident power in this 20 μsec period. Since the distribution of
scattered photons in time is given by a Poisson distribution, a density
measurement with 3% accuracy would thus require 1 kW of laser power
during the 20 μsec. However, at the present state of the art, neither
a 1 kW CW laser, nor a pulsed laser capable of producing 1 kW pulses of
20 μsec duration (or 10 kW for 2 μsec duration, or 100 kW for 200 nsec
duration, since it is the energy content per pulse that matters) at
50 kHz, is available -- at least not in the visible wavelength region.

The situation could be improved somewhat by considering one, or
several of the following points:

a) Accept a 5% accuracy in the density; → gain of a factor of 2.5
b) Increase the focal volume to 1 mm; → gain of a factor of 10
c) Use multiple pass light traps; → gain of a factor of 2→100
d) Increase the sampling time to 40 μsec; → gain of a factor of 4

Application of all of the above will reduce the power required to a
value compatible with present-day lasers. The price one pays is loss of
accuracy both in the actual density, and in the spatial and temporal
resolution.

Another possible mode of attack is to use the entire 1-2 msec pulse
of a ruby laser. At present, there are ruby lasers capable of putting
out 40J over a 1-2 msec pulse duration. Although both the Raman scattering
cross-section and the photomultiplier quantum efficiency are lower at the
longer wavelength of a ruby laser, a 40J pulse for 1.5 msec would on the
average result in 800 photons detected per 20 μsec interval. One drawback with the latter system, however, is the inherent oscillations of the

ruby laser output, with time scales of the order of 1 μsec. So, again, we don't get something for nothing. Nevertheless, this approach appears to be the most promising for the investigation of the density fluctuations in a turbulent shear layer, and preliminary feasibility studies are underway at our laboratories.

The purpose of this discussion was not to give a pessimistic view of the Raman effect as a possible diagnostic tool in fluid dynamics experiments. Rather, it was meant to show that Raman scattering is neither the ultimate answer, nor can one afford to discard it. The experimental situation must be analyzed carefully, and the demands of spatial resolution, frequency response, and desired accuracy must be weighed against the state of the art in laser output power and photodetection efficiency.

It appears that, at least for the present time, the "smallness" of the Raman effect is such that its use is limited in turbulent flow measurements, except at sufficiently high densities. On the other hand, in situations where long time averaging is permissible, the Raman effect may prove vastly superior as a non-interfering density (and temperature) probe, and there are many situations where an optical measurement such as Raman scattering is the only possible diagnostic tool.

REFERENCES

1.   G. L. Brown, and A. Roshko; The Effect of Density Difference on the Turbulent Mixing Layer Turbulent Shear Flows, AGARD-CP-93, Jan 1972 (also to be published in Journal of Fluid Mech.)

2.   M. R. Rebollo, PhD Thesis, California Institute of Technology (1973).

STORM DISCUSSION

PENNEY - If a lower repetition rate, say 1/2 kHz, would produce meaningful results, then a currently available pulsed N_2 laser could be considered. The energy per pulse is about an order of magnitude lower than you would like for your stated accuracy, focal volume, etc., but as you pointed out, if you accept 5% accuracy in the density measurements instead of 3%, and if you increase the focal volume to correspond to a 1 mm spot in the shear layer, then the experiment looks feasible. Also, as you pointed out, multiple passing of light and an increased sampling time could be used.

STORM - Yes, but we do have a strong desire to operate at 10 kHz, that is, one sample every 100 μsec, because these eddies are sweeping by at that rate. However, it is worthwhile to point out that some approach to the ultimately-desired turbulent mixing experiment is conceivable now. Hopefully, new developments in laser technology will come to our aid as well.

HENDRA - Why not find an old Cary 81 instrument. This has a 4-inch high entrance slit on its monochromator. That would mean you could view a long specimen since the virtual image of the slit at the specimen could be at least 2" high. Furthermore, you could get several photomultipliers across the four-inch high exit slit. If you then use red sensitive photo-multipliers and a pulsed ruby laser, I'd guess that you would get something useful.

STORM - But how fast can you pulse the ruby?

HENDRA - I think you would be lucky to get 6 - 8 shots in your two-second run.

STORM - But that is far less than required for a 10kHz sampling rate.

HENDRA - Maybe, but wouldn't it be a significant indication to you if you got simultaneous information at, say, six points across your sample zone?

STORM - It would be some help, yes. But we already have calibrated density probes available at Caltech that could sweep across the flow in less than a second, and these probes have responses to 10kHz. Of course we would prefer a non-interfering probe.

MOREY - There is a device called a Reticon that is an array of photodiodes that are spaced on approximately 1 mil centers, and the whole sensitive area is about 1cm long. It might be possible to image a laser beam onto this, giving more spatial resolution.

SCHILDKRAUT - There is more noise current than 140 photons worth in an individual diode array element. Since there is no gain in the chip, the signal would be swamped by bulk semi-conductor noise.

LAPP - Since we are talking about mixing nitrogen into helium, you can use interference filters. These would allow large apertures and collection angles. Furthermore, you could consider using a multipassing configuration for the incident laser beam.

STORM - That might be barely feasible for this application, but it is too close for comfort.

PENNEY - Could you use Rayleigh scattering? That would differentiate between nitrogen and helium, because helium has a very small Rayleigh cross section, about 1/20 of that of nitrogen.

STORM - Possibly, but then one encounters the potential interference of scattering from particles.

PENNEY - True, but particles larger than perhaps 0.1 micron will produce pulses that can be subtracted out.

TARAN - Are all the feasibility calculations which you made based upon vibrational scattering?

STORM - Yes. I realize that the cross section for integrated rotational scattering is over an order of magnitude larger, which would make the experimental estimates more optimistic for situations where we can use rotational scattering.

SCHILDKRAUT - I just want to caution you about one thing since I have raised the question of an ellipsoidal collector at this meeting. The depth of field of a large aperture ellipsoid (f 0.3) which collects over 2 steradians is as low as 1500 Å in dimension. The sample size (diameter) you will be looking at, trying to image through a practical system, may go as "high" as 10 microns. This is incredibly small. What depth are you interested in sampling?

STORM - As it turns out, small depths are worthwhile here, since the flow is strongly two dimensional.

SCHILDKRAUT - You have to worry about how many molecules are in the volume you are looking at, that is the only thing I want to caution you about. Few molecules multiplied by small cross sections results in very few Raman photons. Two steradians is O.K. but there is not much sample volume imaged.

WHITE - One could use a Q-switched ruby laser, and cause the beam to make multiple traverses of the flow, each traverse displaced in space from the preceding one. By having the mirrors several feet from the flow, the external path length can be such that the scattered light from successive traverses does not overlap in time. One photomultiplier then may be used to sense the scattered intensity from selected regions including multiple traverses, and the spatial location of the scattering source determined from the time delay in the scattered light.

SCHILDKRAUT - That is a practical idea; it works! That kind of vertical scanning is an extremely potent way to utilize the laser energy if the medium is not highly absorbent. Supposing each pass between mirrors is 2-3 feet; that's only a few nanoseconds and is within the time resolution capabilities of the new PMT's. Placing staggered mirrors to arrange the zig zag of a well collimated laser beam is not at all difficult and you wouldn't loose too much.

STORM - Thank you all for your suggestions.

FOURIER TRANSFORM RAMAN SPECTROSCOPY

by

E. Robert Schildkraut
Tomas B. Hirschfeld

Block Engineering, Inc.
19 Blackstone Street
Cambridge, Massachusetts

ABSTRACT

The application of Fourier Transform Spectrometers to Laser Raman Spectroscopy is briefly discussed. Fundamental limitations of the technique, its applications, and prognosis for the near future are noted. References which relate to calculations on input Rayleigh filtration requirements are listed.

* * * * * * * *

With the coming of age of Fourier Transform Spectroscopy,[1,2] and the increased number of applications for Raman spectroscopy, a few words on the potential (or lack thereof) for combining these two powerful techniques are in order. To begin with, let us restrict this brief discussion to cw laser Raman spectroscopy. The use of pulsed lasers and pulse gated detectors, while not basically incompatible with Fourier Transform Spectroscopy, does tend to lose some of the multiplexing advantage,[3] and certainly complicates the interferogram which ultimately yields the spectrum.

FTS used in a Raman system, because of the large throughput[4] obtained due to the absence of spectrometer slits, could reduce considerably the time needed to obtain a spectrum. Likewise, the multiplex advantage[5] obtained with some types of detectors by looking at all of the wavelengths all of the time could, in principle, make even a typical weak Raman spectrum obtainable in less than one minute. This is particularly true for IR Raman where PMT's and other photon noise limited detectors are not used.[6]

The problems associated with FTS/Raman as a general tool stem from the noise in the laser source and the particular way in which noise becomes redistributed in the Fourier transform process.

In traditional laser Raman spectroscopy, double, triple[7] and even
quadruple cascaded monochromators are used to separate or attenuate the
elastically scattered laser light from the Raman spectral information.
In a scanning spectrometer, the residual laser light falls off to a great
extent as one scans away from the laser wavelength ($\Delta\nu = 0$).[8] Noise in
the laser line or background is thus well restricted to the laser wavelength alone and, of course, the Raman spectral peaks which track along
with it. Between the peaks we have low background and also low noise.

The situation is different in FTS. Picture a purely monochromatic
laser with a very low, but obtainable 1% amplitude[9] noise over a broad
audio frequency spectrum. Such a "delta function" spectral spike is
seen by an interferometer (Figure 1) as a pure sinusoid with noise

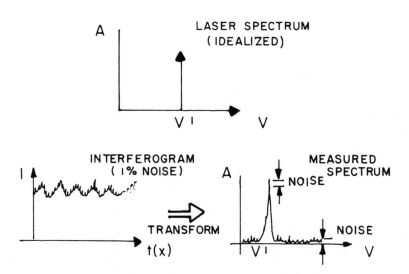

Figure 1. Laser spectrum in frequency and time domains.

uniformly distributed over time. The amplitude of the laser varies
randomly as the interferometer makes its scan and this adds to the
modulation produced by the interference phenomenon (fringes). When this
noisy sinusoid is transformed by an interferometer, all of the noise
remains uniformly distributed over the entire spectral plot.[10] In contrast, a dispersive spectrometer tuned away from the laser line would
see little or no laser generated noise (Figure 2).

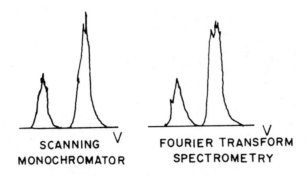

Figure 2.   Laser-generated noise seen by scanning monochromator
            and Fourier-transform spectrometer.

Now if we look at a Raman spectrum which consists of only one narrow
Raman line (for simplicity), our interferogram would have the appearance
of two added sinusoids (at different frequencies) both containing uniformly distributed noise (Figure 3). Of course, the Raman shifted
sinusoid would be much lower in amplitude than as shown exaggerated
in Figure 3.

When we finally put in the ratio of Raman/Rayleigh intensities for
typical solids and liquids, the noise RMS amplitude tends to swamp the
weak Raman signal from vibrational rotational lines. Electronically
filtering the noise from the detector signal prior to transforming only
serves to degrade spectral resolution, since the noise and signal audio
spectra overlap. The problem is then to filter out the elastically
scattered radiation (Rayleigh, Mie, or other) before it enters or is
processed by the interferometer.[11] Calculations show that spike filter
attenuation of from 10^6 to 10^{12} is needed before FTS Raman becomes
practical in general.[12] For certain applications of course, a 10^3 laser
line attenuation might even be adequate, but this is not the general case.
Pure rotational Raman spectroscopy requires working very close to the
laser wavelength and this tends to cancel the advantage of added signal
intensity. Spike interference filters have low transmission and limited
angular acceptance if their narrow bandwidth is to be preserved. Absorption
filters for the popular excitation laser wavelengths, where they exist,
are not ready for commercial implementation as yet.[13]

By modulating the detector or amplifier gain of an interferometer
at an audiofrequency ω_x, and maintaining a proper phase relationship
with the mirror scan, the spectral function of the instrument can be
made to develop a gap at ν_x. Achieving the same object by putting an
electro-optical modulator in front of the interferometer eliminates part
of any noise carried in the input beam itself at ν_x. This might be useful
in order to reduce the shot noise carried by ν_x.

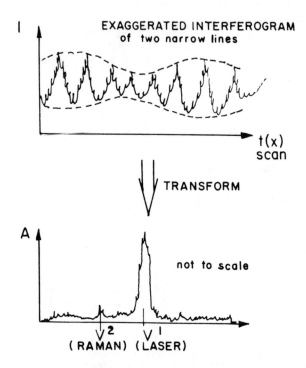

Figure 3.   Schematic representation of waveforms from a Fourier
            transform spectrometer used for Raman spectroscopy.

Modulation efficiencies <100% will limit the extent to which ν_x's
shot noise or other amplitude noise can be suppressed in the electro-
optical input modulator, and provide a gray level in the synthetic source.
But there are other limits to the efficiency of such systems.

Cascading filters to achieve higher laser attenuation reduces
transmission in the passband to less than 1%[14] and this signal loss can
obliterate any gain in speed due to FTS.

Far from least important is the choice of detector.   For Raman
spectroscopy in the visible region, photomultiplier tubes are used
almost exclusively. Electron multiplication with the newer high gain
tubes is practically noise free and photon or shot noise is the limiting

factor for such detectors. Photon noise is proportional to $\sqrt{b}$ where b is the number of photons arriving per second. The interferometer transmits roughly N/2*times as much light[15] as a dispersive spectrometer with slits, hence the $\sqrt{N/2}$ Felgett's advantage of interferometers is exactly cancelled by the increased photon noise in cases where a uniformly distributed spectrum (continuum) is present. When only a few lines appear, the situation may be statistically different but not enough to eliminate this problem. In infrared Raman spectroscopy where shot noise is well below the detector noise, Felgett's advantage is still relevant.

Finally we come to a temporary difficulty - fabrication. Interferometer mirrors (for use in the visible region) of adequate flatness and size have been and can be made; maintaining parallelism over an entire high resolution scan is a challenge for present bearing configurations and quality. The angular tilt allowable in a scanning mirror used at 4880 Å for example is several times less than can be tolerated in a near IR interferometer.

In summary then, the problems inherently associated with FTS/Raman make it an unwise choice for practical, general purpose laser Raman spectroscopy in the visible. Future work on spike attenuation filters may increase the potential of this technique in certain restricted applications.

REFERENCES

1.   Robert John Bell, Introductory Fourier Transform Spectroscopy, (Academic Press, New York, 1972), pp. 1-15, 193-195.

2.   Marvin T. Margoshes, "Air Pollution Monitoring," Reprint available from Digilab, Inc., 237 Putnam Ave., Cambridge, Mass. 02139.

3.   Lawrence Mertz, Transformations in Optics, (John Wiley and Sons, New York, 1965), p. 7-8.

4.   Ibid. p. 15.

5.   Ibid. p. 8.

6.   Ibid. p. 8-9.

7.   Cary Instruments Inc. Bulletin 182 (June 1970).

8.   J. F. James, and R. S. Sternberg, The Design of Optical Spectrometers, (Chapman & Hall, London, 1969), p. 7-8.

9.   Coherent Radiation Inc. Bulletin 52G.

* N here is the number of resolution elements within the pathband of the interferometer.

10. Op. Cit., Mertz, p. 10.

11. R. B. Blackman and J. W. Tukey, The Measurement of Power Spectra,
     (Dover Press, New York, 1958), p. 28-29.

12. T. H. Hirschfeld and P. R. Griffiths, "A Comparison Between
     Fourier Transform Spectroscopy and Conventional Scanning Techniques,"
     Columbus Symp. Mol. Struct. and Spectry., Columbus, Ohio, (September
     1969).

13. H. H. Claassen, Applied Spectroscopy, 23, 8 (1969).

14. S. H. Melfi, et al, "Observation of Raman Scattering by SO_2
     in a Generating Plant Stack Plume," Appl. Phys. Lett. 22, 402 (1973).

15. Op. Cit., Mertz., p. 9.

SCHILDKRAUT DISCUSSION

LEONARD - Suppose you filter out the laser line and then do the
transform.

SCHILDKRAUT - That is the very point. If you can get a filter that
is good enough to match the throughput of the interferometer, and block
the laser line, then Raman Fourier transform interferometry is a practical thing to do. We haven't got that filter yet.

LAPP - What characteristics do you need?

SCHILDKRAUT - We would like to have approximately 10^{10} to 10^{12}
attenuation of the laser and have roughly an f/4 cone through that filter.

LAPP - I'm not sure I understand what you mean by the filtering.
Could you describe exactly what the characteristics should be?

SCHILDKRAUT - Yes, we wish a transmission notch at the laser wavelength. It would be acceptable to pass one side or the other side of
the band also if I want a high or low pass filter, but we must eliminate
the laser photons before they get to the interferometer. The filter
should accept a f/4 cone of light and still do that much filtering.

LEONARD - How good is that plastic that you have used for
filtering?

SCHILDKRAUT - Not that good. I think the best it has done for us
is 10^4 - 10^6, at 3472 Å. I don't have a plastic for 4880 attenuation.

LEONARD - If you make that sheet twice as thick, do you get 10^{12}?

SCHILDKRAUT - For absorption filters yes, but the cut-off wavelength also shifts. What happens when you start doing that is you start cutting into Raman region as well, since this is a pure absorption.

LAPP - Is there any way of lessening the severity of this problem by stabilizing the frequency of the laser?

SCHILDKRAUT - Yes, to some extent, there are some things that can be done that haven't been perfected as yet. I imagine there is almost sufficient interest to start doing them but you may run into plasma stability problems, and I am far from an expert in that area; I don't know what the limit is. The photon noise probably is the ultimate limit, in other words, we could get down to the point where the laser is stable compared to the photon noise. The frequency and amplitude stability of the laser are not that tightly related and it is amplitude stability which hurts us here.

LAPP - Are feedback processes applied to these lasers good enough to help solve these problems?

SCHILDKRAUT - The noise contains some pretty high frequency components; the feedback mechanisms are much slower than that. The servos for current are 1 kilo-hertz servos, typically.

BARRETT - What happens if you just reduce the laser intensity to about the magnitude of the Raman lines?

SCHILDKRAUT - That is probably sufficient to see the Raman lines but the laser noise will still be noticeable.

BARRETT - With the filter, why do you have to go 10^{12}? Because if you have flicker in the laser, you are going to have flicker on your Raman spectrum.

SCHILDKRAUT - The 10^{12} number I quoted is a rough guess at a desirable value. How many Raman photons compared to Rayleigh photons do you collect? If you bring the raw amplitude of both signals into line, the filtering is adequate.

BARRETT - But you still have the noise on the Raman line, right?

SCHILDKRAUT - Exactly, I am trying to figure out what you really need to make this into a usable device or product.

BARRETT - Perhaps you ought to engineer a superstable laser in which you eliminate the noise.

SCHILDKRAUT - I don't know what is involved in engineering the laser to be superstable. We really do fairly well now; we used to get 5% noise specs; I really don't know what is involved to get it down to 10^{-5}.

TARAN - Couldn't you suppress that noise from the response of the detector with a low pass filter in the electronics, while scanning the interferometer slowly?

SCHILDKRAUT - No, the noise is broadband and exists every where the Raman information exists. A very low pass filter severely degrades the resolution of the instrument and the lines disappear.

MACK - How about atomic resonance filters, such as one based on sodium vapor absorption?

SCHILDKRAUT - Well, we were looking at rare earth glass among other things, but it is true the absorption filters hold reasonable promise where they exist.

MACK - Well, the transmission passband for that material would be quite broad compared to that for sodium vapor.

SCHILDKRAUT - And then the stability in the sodium vapor absorption cell is open to question. I don't know typically what that does but it could add noise due to its own modulation.

MACK - I don't think you get enough through to see anything. The cross section for sodium vapor on line center is perhaps 10^{-10} cm^2.

BARRETT - Have you actually tried to run a spectrum?

SCHILDKRAUT - Yes, I have tried a crude experiment because I didn't believe it at first.

BARRETT - And what happened?

SCHILDKRAUT - The noise does in fact transform exactly as is expected. The power spectral density of the noise in the laser line transforms to the same power spectrum spread out over wavelength space.

LEONARD - Suppose you are trying to see a small Raman component against a large Raman component and you can't filter one without the other?

SCHILDKRAUT - That is a secondary problem, but it is similar in the sense that you would have a large noise in the strong component. I would like to get to the point where I see that trouble; it's several orders of magnitude less trouble than the laser noise.

LEONARD - I am thinking about the problems I have with the small concentration pollutants when you also have the nitrogen Raman present as a large signal.

SCHILDKRAUT - The shot noise in the nitrogen line is going to tend to obliterate the signal from the lower concentration constituent. As I said, I would really like to get to the point where I would see that in the experiment and have to worry about it. The FTS system would suffer worse than your scanning monochromator in the CO/N_2 discrimination problem.

BARRETT - What you are saying then, is mainly it is the noise in the laser.

SCHILDKRAUT - Yes.

BARRETT - So you make the filter at the laser line, but you still have bandpass at your Raman lines. You're going to be working on a different level now.

SCHILDKRAUT - Right. You are working in a different sensitivity region and now the throughput advantage for certain experiments makes the system worthwhile to explore. This CO vs. nitrogen wing problem is very severe. There are other problems not as severe where you could get the spectrum in a fraction of a second. In all this optical ginger-bread what you are really trying to do is gather as much light from a reasonably large sample volume as possible.

TARAN - Now, going back to compressing the spectrum, couldn't you filter out the noise fluctuations with a narrow band filter in the detector output? If the interferometer is scanned at a fixed rate, the near monochromatic light coming out of it will be modulated in amplitude at a fixed, near monochromatic, frequency; pick that particular frequency in the detector output, and reject the other components outside a certain bandwidth with a narrow band filter.

SCHILDKRAUT - Are you talking about electrically filtering the signal now?

TARAN - Yes.

SCHILDKRAUT - No, the noise in the laser line is what hurts, not the monochromatic laser line itself. There is no filter for the noise; it's broadband. In addition, the practical types of electrical filters that are available at state of the art, like 40 db per octave, 80 db per octave, are not satisfactory; I'm talking about 10^6 here. If you cut sharp enough, phase distortion comes into play. If you try to go 600 db per octave, then the phase distortion that you get out of this device is going to make your spectrum unrecognizable. Digital filtering might be possible.

TARAN - Yes, but the phase distortion might just shift the line slightly or change its shape, which is not so serious, is it?

SCHILDKRAUT - The phase can be extremely serious in the inverse transform. Some of the lines can go negative. That is pretty disturbing!

? - Just square it!

SCHILDKRAUT - That's called a pure amplitude transform. It is not good enough for most spectroscopists and does not solve the problem of line shift and overlap.

DELHAYE - In addition, you have to know which frequency to filter before you obtain an unknown spectrum.

TARAN - The line shift and distortion, though, would not be so serious if you are only interested in concentration measurements.

SCHILDKRAUT - I think it would. I should have prefaced this by saying that I tend to think in terms not necessarily of a scientific experiment for a specific application, but devices that have broad use, and I really don't see that the latter is practical yet with the tools we have; I'm sure that for certain experiments Raman Fourier transform interferometry might still be a very valuable technique, and I should have made that clear. I'm not negating any specific experiment which provides a tool for the Raman spectroscopist.

LAPP - You are looking toward the time analogous to that for the introduction of Fourier Transform infrared instruments.

SCHILDKRAUT - Yes. The dedicated mini-computer made that possible and practical or we might as well have packed up, and not even tried it.

LIST OF ATTENDEES

Dr. J. J. Barrett
Materials Research Center
Allied Chemical Corporation
P. O. Box 1021R
Morristown, New Jersey 07960

Professor Daniel Bershader
Department of Aeronautics
   and Astronautics
Stanford University
Stanford, California 94305

Dr. George Bethke
Room L9517
G.E. Space Science Lab
P. O. Box 8555
Philadelphia, Pennsylvania 19101

Dr. Graham Black
Stanford Research Institute
Menlo Park,
California 94025

Dr. Anthony Boiarski
Code 313, Aerophysics Branch
Naval Ordnance Laboratory
White Oak
Silver Spring, Maryland 20910

Lt. Gerald E. Bresowar
AFAPL/SFF
Wright-Patterson Air Force Base,
Ohio 45433

Mr. Donald L. Champagne
Gas Turbine Engineering
General Electric Company
Schenectady, New York 12345

Major Maurice Clermont
AEDC/DYR
Arnold Air Force Station,
Tennesses 37309

Dr. John W. Daiber
Calspan Corporation
4455 Genesee Street
Buffalo, New York 14221

Mr. F. L. Daum
ARL/LF
Wright-Patterson Air Force Base,
Ohio 45433

Professor M. Delhaye
Universite des Sciences et Techniques
   de Lille
Laboratoire de Spectroscopie Raman
Centre de Spectrochimie - B.P. 36
59650 Villeneuve D'ASCQ, France

Mr. Lee Dodge
United Aircraft Research Labs
United Aircraft Corporation
400 E. Main Street
East Hartford, Connecticut 06108

Dr. E. Stokes Fishburn
A.R.A.P.
50 Washington Road
Princeton, New Jersey 08540

Professor R. Gaufres
Laboratoire de Spectroscopie Moleculaire
Universite des Sciences et Techniques
   du Languedoc
Place E. Bataillon
34 - Montpellier, France

Dr. A. S. Gilbert
Division of Pure Chemistry
National Research Council of Canada
Ottawa, Canada

Dr. Robert Goulard, Director
Project SQUID
Thermal Science and Propulsion Center
Purdue University
Lafayette, Indiana 47907

Dr. D. L. Hartley
Sandia Laboratories
Division 8364
P. O. Box 696
Livermore, California 94550

Dr. A. B. Harvey
Naval Research Laboratory
Code 6110
Washington, D.C. 20375

Dr. William H. Heiser
Chief Scientist
AFAPL/SFF
Wright-Patterson Air Force Base,
Ohio 45433

Professor P. J. Hendra
Department of Chemistry
University of Southampton
Southampton SO9 5NH, England

Dr. Ronald A. Hill
Sandia Laboratories
Division 5642
Albuquerque, New Mexico 87115

Dr. W. W. Hunter, Jr.
Gas Parameters Measurements Section
Measurements Physics Branch
NASA Langley Research Center
Hampton, Virginia 23365

Dr. Wolfgang Kiefer
Sektion Physik der Universitat
Lehrstuhl Prof. Dr. J. Brandmuller
Amalienstrasse 54/IV
8000 Munchen 13, West Germany

Dr. Marshall Lapp
General Electric Company
Corporate Research and Development
P. O. Box 8
Schenectady, New York 12301

Professor Samuel Lederman
Department of Aerospace Engineering
   and Applied Mechanics
Polytechnic Institute of New York
Farmingdale, New York 11735

Dr. Donald A. Leonard
AVCO Everett Research Laboratory
2385 Revere Beach Parkway
Everett, Massachusetts 02149

Dr. J. W. L. Lewis
ARO Inc. - VKR/ADP
AEDC
Arnold Air Force Station,
Tennessee 37389

Dr. Yung Liu
General Electric Company
Corporate Research and Development
P. O. Box 8
Schenectady, New York 12301

Dr. Michael E. Mack
AVCO Everett Research Laboratory
2385 Revere Beach Parkway
Everett, Massachusetts 02149

Dr. S. Harvey Melfi
EPA NERC - LV
P. O. Box 15027
Las Vegas, Nevada 89114

Dr. William Morey
General Electric Company
Corporate Research and Development
P. O. Box 8
Schenectady, New York 12301

Mr. Paul Mossey
Measurement Development
Building 301, Mail Drop H-76
Evendale Plant
General Electric Company
Cincinnati, Ohio 45215

Mr. David Murray
ARL/LF
Wright-Patterson Air Force Base,
Ohio 45433

Dr. G. B. Northam
Environmental and Space Sciences Div.
NASA Langley Research Center
Building 1230, Mail Stop 234
Hampton, Virginia 23365

Mr. James R. Patton, Jr.
Office of Naval Research
Power Program, Code 473
Department of the Navy
Arlington, Virginia 22217

Dr. C. M. Penney
General Electric Company
Corporate Research and Development
P. O. Box 8
Schenectady, New York 12301

Dr. Arthur V. Phelps
JILA Building, Room 800
Joint Institute for Laboratory
    Astrophysics
University of Colorado
Boulder, Colorado 80302

Dr. Ralph Roberts
Office of Naval Research
Director, Power Program, Code 473
Department of the Navy
Arlington, Virginia 22217

Mr. J. A. Salzman
NASA Lewis Research Center
Mail Stop 54-3
21000 Brookpark Road
Cleveland, Ohio 44135

Dr. E. Robert Schildkraut
Director of Research
Block Engineering, Inc.
19 Blackstone Street
Cambridge, Massachusetts 02139

Dr. Robert C. Sepucha
Aerodyne Research, Inc.
Northwest Industrial Park
Burlington, Massachusetts 01803

Dr. Robert Setchell
Sandia Laboratories
Division 8364
P. O. Box 696
Livermore, California 94550

Mr. J. R. Smith
Sandia Laboratories
Division 8333
P. O. Box 696
Livermore, California 94550

Dr. David A. Stephenson
Physics Department
General Motors Research Laboratories
12 Mile and Mound Road
Warren, Michigan 48090

Dr. Eric Storm
California Institute of Technology
1201 E. California Boulevard
Pasadena, California 91109

Dr. J. -P. E. Taran
Office National d'Etudes et de
    Recherches Aérospatiales (ONERA)
92320 Chatillon, France

Dr. Donald R. White
General Electric Company
Corporate Research and Development
P. O. Box 8
Schenectady, New York 12301

AUTHOR INDEX

393

This index is divided into major and minor categories. The division is based primarily upon usefulness of the information for applications. Thus, "wind tunnel" is found under "Measurement environment," while, on the other hand, "Atmospheric transmission" is given as a major category. Where extensive discussion of a listed topic is to be found on a given page as well as on succeeding pages (or perhaps throughout the article), the page number is underlined.